★ ★ ★ ★ ★ ★ ★ ★ ★ ★ ★

BARNES & NOBLE

is a proud sponsor and exclusive retail partner of the

FLAGSHIP MAKER FAIRES

BARNES & NOBLE
Mini Maker Faire

Our year of sponsorship culminates in our
2ND ANNUAL MINI MAKER FAIRE
in all stores nationwide Nov 5 + Nov 6
FOR MORE INFORMATION **VISIT BN.COM/MAKERFAIRE**

CONTENTS

Make: Volume 51
June/July 2016

COLUMNS

Reader Input 08
Thoughts, tips, and musings from readers like you.

Welcome: Flying into the Future 10
Makers are raising autonomous aircraft to new heights.

Made on Earth 12
Backyard builds from around the globe.

FEATURES

Maker ProFile: Crowdfunding Coach 16
Helpful advice from Kickstarter's Senior Director of Design and Technology Communities John Dimatos.

ON THE COVER:
Drones take flight. Know the inspiration for this image? Tell us and we'll give you a shout-out. editor@makezine.com Illustration: Raul Arias

Special Section 18

DRONE REVOLUTION

Aerial Aid 20
Amateur drone pilots can save lives, find bodies — or wreak logistical and legal havoc.

Little Dipper Drone 26
Learn working knowledge of aerial robotics by building this 300-class autonomous quadcopter.

2016 Drone Flyer's Guide 31
We review 8 capable quadcopters and 8 DIY-drone flight controllers to help get you airborne.

FPV Night Flying 38
Hack an infrared flood lamp and rip through the darkness!

Top Guns 40
These multirotor vanguards are pushing the field to new heights.

Build Your Own Drone 42
Get aerial without breaking the bank by making these quadcopters and fixed-wing R/C projects.

Anti-Drone Wi-Fi Hijacker 44
Build a Pi-powered drone disabler to understand the security risks of wireless communications.

SKILL BUILDER

Servos 101 48
Learn to make things move with this simple motor.

Polish with Wet Sanding 50
This process removes large scratches left during shaping.

makershed.com

makezine.com/51

PROJECTS

Get Your Freq On 52
Construct a music visualizer table that dances to your tunes!

Custom Catwalk 58
This light-duty floating shelf hangs with no visible support.

1+2+3: Rhythm Bones 59
Make your own set of traditional bone instruments.

Tot-Sized Tank 60
Tackle any terrain in this nigh-unstoppable tracked vehicle.

Remaking History: Chester Rice and the Dynamic Loudspeaker 66
Build the moving-coil, direct-radiation transducer that has rocked our world for 90 years.

3D-Printed Tourbillon Clock 68
Print and assemble this working, large-scale model of a precision watch design.

Percusso: MIDI-Controlled Percussion Tower 72
How one Maker created his crazy rhythm bot with simple solenoids and a MIDIWidget.

Ponytrap: An Arduino Robot Drummer 75
Build a hard-charging drumbot with real drumsticks.

Mario Play Cubes 76
Turn vintage video game pixels into an easy macramé pattern.

DIY Smart Light Switch 78
Learn to use Bluetooth to control AC lamps and devices from your smartphone.

Giant Vortex Air Cannon 82
Bring the boom with supersized smoke rings that go the distance.

Amateur Scientist: Ultra-Simple Solar Radiometer 84
No batteries, no switches — this analog meter is powered by the sunshine it's measuring.

1+2+3: Wearable EL Flame 86
Craft glowing EL panels into eye-catching accessories.

TOOLBOX

Tool Reviews 88
DeWalt orbit sander, Black & Decker cordless glue gun, Tekton 16-piece screwdriver set, and more useful gear.

3D Printer Review: Ultimaker 2+ 92
A new and improved extruder produces crisp, clean, and impressive prints.

OVER THE TOP

Over the Top: Crazy Train 96
Ozzy Osbourne joins the Gremlins Carnival Club.

Issue No. 51, June/July 2016. *Make:* (ISSN 1556-2336) is published bimonthly by Maker Media, Inc. in the months of January, March, May, July, September, and November. Maker Media is located at 1160 Battery Street, Suite 125, San Francisco, CA 94111, 877-306-6253. SUBSCRIPTIONS: Send all subscription requests to *Make:*, P.O. Box 17046, North Hollywood, CA 91615-9588 or subscribe online at makezine.com/offer or via phone at (866) 289-8847 (U.S. and Canada); all other countries call (818) 487-2037. Subscriptions are available for $34.99 for 1 year (6 issues) in the United States; in Canada: $39.99 USD; all other countries: $50.09 USD. Periodicals Postage Paid at San Francisco, CA, and at additional mailing offices. POSTMASTER: Send address changes to *Make:*, P.O. Box 17046, North Hollywood, CA 91615-9588. Canada Post Publications Mail Agreement Number 41129568. CANADA POSTMASTER: Send address changes to: Maker Media, PO Box 456, Niagara Falls, ON L2E 6V2

THE FUTURE IS MADE HERE

JULY 15-16, 2016
AT&T PARK
SAN FRANCISCO

Imagine what's next with us. Robots as our best friends. Cities as our jungle gyms. Together with Maker Camp, you'll get to understand and interact with the technologies that will shape our future.

shape.att.com

Make:

EXECUTIVE CHAIRMAN & CEO
Dale Dougherty
dale@makermedia.com

CFO
Todd Sotkiewicz
todd@makermedia.com

VICE PRESIDENT
Sherry Huss
sherry@makermedia.com

> "'The Guide says there is an art to flying,' said Ford, 'or rather a knack. The knack lies in learning how to throw yourself at the ground and miss.'" — *Douglas Adams, from Life, the Universe, and Everything*

EDITORIAL

EXECUTIVE EDITOR
Mike Senese
mike@makermedia.com

DIRECTOR OF CONTENT & COMMUNITY
Will Chase
willchase@makermedia.com

PRODUCTION MANAGER
Craig Couden

PROJECTS EDITORS
Keith Hammond
khammond@makermedia.com
Donald Bell

SENIOR EDITOR
Caleb Kraft
caleb@makermedia.com

ASSISTANT EDITOR
Sophia Smith

COPY EDITOR
Laurie Barton

EDITORIAL INTERN
Lisa Martin

CONTRIBUTING EDITORS
Stuart Deutsch
William Gurstelle
Nick Normal
Charles Platt
Matt Stultz

CONTRIBUTING WRITERS
Alasdair Allan, Phil Bowie, Jordan Bunker, Brian Bunnell, Brent Chapman, Don Coleman, Larry Cotton, DC Denison, Helga Hansen, Tom Heck, Michelle Hlubinka, Patty Hodapp, Justin Kelly, Belinda Kilby, Terry Kilby, Christophe Laimer, Dan Maxey, Forrest Mims III, Sandeep Mistry, Jane Stewart, Quentin Thomas-Oliver, Charlie Turner, Alex Zvada

DESIGN, PHOTOGRAPHY & VIDEO

ART DIRECTOR
Juliann Brown

DESIGNER
James Burke

PHOTO EDITOR
Hep Svadja

SENIOR VIDEO PRODUCER
Tyler Winegarner

MAKER MEDIA LAB

LAB COORDINATOR
Emily Coker

LAB INTERNS
Anthony Lam
Jenny Ching

MAKEZINE.COM

DESIGN TEAM
Eric Argel
Beate Fritsch

WEB DEVELOPMENT TEAM
David Beauchamp
Rich Haynie
Loren Johnson
Bill Olson
Ben Sanders
Clair Whitmer
Alicia Williams
Wesley Wilson

CONTRIBUTING ARTISTS
Raul Arias, Matthew Billington, Rob Nance, Andrew J. Nilsen, Damien Scogin, Peter Strain

ONLINE CONTRIBUTORS
Raphael Abrams, Cabe Atwell, Gareth Branwyn, Jon Christian, Ian Cole, Jeremy Cook, Kathy Ceceri, Marc de Vinck, Jimmy DiResta, Michael Floyd, Bilal Ghalib, Paul Gentile, Will Holman, Patrick Houston, Peter Konig, Jeanne Loveland, Russell Munro, Luanga Nuwame, Darbin Orvar, Pete Prodoehl, Sean Michael Ragan, Andrew Salomone, Daniel Schneiderman, Nicholas Squires, Andrew Terranova, Bill Tomiyasu

SALES & ADVERTISING
makermedia.com/contact-sales or sales@makezine.com

SENIOR SALES MANAGER
Katie D. Kunde

SALES MANAGERS
Cecily Benzon
Brigitte Kunde

STRATEGIC PARTNERSHIPS
Angela Ames
Allison Davis

CLIENT SERVICES MANAGER
Mara Lincoln

MARKETING

CHIEF MARKETING OFFICER
Amy Maniatis
amy@makermedia.com

MARKETING COMMUNICATIONS MANAGER
Brita Muller
brita@makermedia.com

MARKETING SALES DEVELOPMENT MANAGER
Jahan Djalali

DIGITAL COMMUNICATIONS SPECIALIST
Kathryn Lastufka

BOOKS

PUBLISHER
Roger Stewart

EDITOR
Patrick Di Justo

MAKER FAIRE

PRODUCER
Louise Glasgow

PROGRAM DIRECTOR
Sabrina Merlo

MARKETING & PR
Bridgette Vanderlaan

SPONSOR RELATIONS MANAGER
Miranda Mota
miranda@makermedia.com

COMMERCE

GENERAL MANAGER OF COMMERCE
Sonia Wong

SENIOR BUYER
Audrey Donaldson

E-COMMERCE MANAGER
Michele Van Ruiten

INVENTORY PLANNER
Percy Young

CUSTOMER SERVICE

CUSTOMER SERVICE REPRESENTATIVE
Jay Johnston

Manage your account online, including change of address:
makezine.com/account
866-289-8847 toll-free in U.S. and Canada
818-487-2037,
5 a.m.–5 p.m., PST
cs@readerservices.makezine.com

PUBLISHED BY
MAKER MEDIA, INC.
Dale Dougherty

Copyright © 2016 Maker Media, Inc. All rights reserved. Reproduction without permission is prohibited. Printed in the USA by Schumann Printers, Inc.

Comments may be sent to:
editor@makezine.com

Visit us online:
makezine.com

Follow us:
@make @makerfaire @makershed
google.com/+make
makemagazine
makemagazine
makemagazine

CONTRIBUTORS

If you could set anything in your house, electronic or not, to "person-detection" mode, what would it be and why?

Patty Hodapp
Minturn, Colorado (Aerial Aid)
I'd set the puppy pooper scooper to doggy-detection mode, well for obvious reasons. It'd make the not so great task of cleaning up after the dogs outdoors a hell of a lot easier.

Anthony Lam
San Francisco, Calif (Test builder, Dynamic Loudspeaker)
I wish my front door would have "person-detection" mode. If it were capable of recognizing my face, it would unlock the door for me without using my key whenever I return home.

Charlie Turner
Milton Keynes, U.K. (Music Visualizer Table)
At night, I want the skirting board in my house to light up the 6 feet around me so I can find my way to the bathroom in the dark!

Christophe Laimer
Hedingen, Switzerland (Tourbillon Clock)
I'd probably use it for fun, e.g. a cyber-pet. This pet could act like an autonomous mobile phone: If somebody calls me, the pet searches for me and impersonates the caller.

Jane Stewart
Guildford, U.K. (Super Mario Play Cubes)
How about a fly drone so you could literally have a fly on the wall and find out who finished the milk, where the toddler hid your keys, and what the dog did to your slippers.

PLEASE NOTE: Technology, the laws, and limitations imposed by manufacturers and content owners are constantly changing. Thus, some of the projects described may not work, may be inconsistent with current laws or user agreements, or may damage or adversely affect some equipment. Your safety is your own responsibility, including proper use of equipment and safety gear, and determining whether you have adequate skill and experience. Power tools, electricity, and other resources used for these projects are dangerous, unless used properly and with adequate precautions, including safety gear. Some illustrative photos do not depict safety precautions or equipment, in order to show the project steps more clearly. These projects are not intended for use by children. Use of the instructions and suggestions in *Make:* is at your own risk. Maker Media, Inc., disclaims all responsibility for any resulting damage, injury, or expense. It is your responsibility to make sure that your activities comply with applicable laws, including copyright.

Make bikes learn new tricks.

What if the act of transporting vaccines could also keep them at the right temperature?

Anurudh, 15
2015 Global Finalist

Everything's better with science.
www.googlesciencefair.com

Google Science Fair

Google · LEGO education · NATIONAL GEOGRAPHIC · SCIENTIFIC AMERICAN · VIRGIN GALACTIC

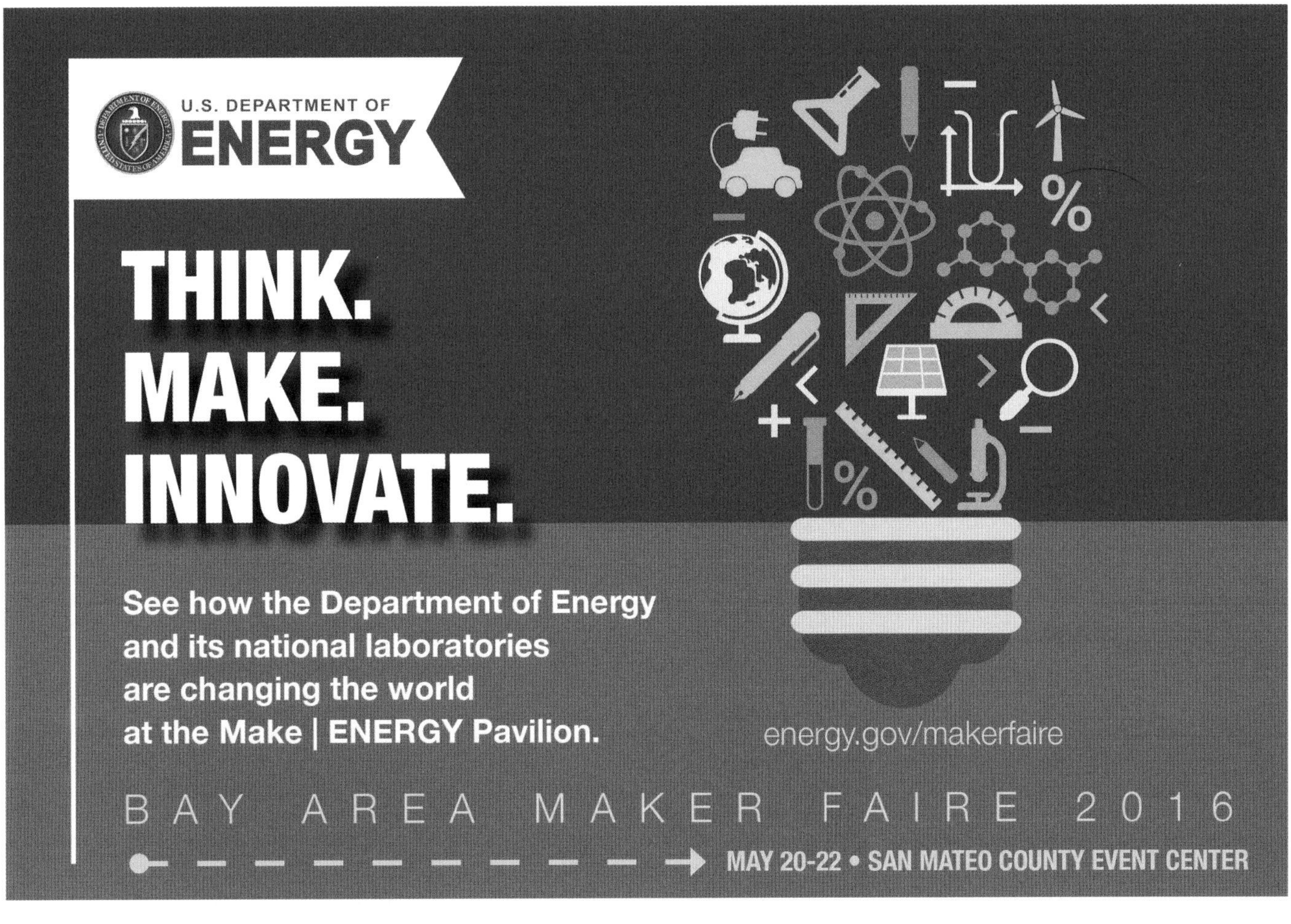

U.S. DEPARTMENT OF ENERGY

THINK.
MAKE.
INNOVATE.

See how the Department of Energy and its national laboratories are changing the world at the Make | ENERGY Pavilion.

energy.gov/makerfaire

BAY AREA MAKER FAIRE 2016

MAY 20-22 • SAN MATEO COUNTY EVENT CENTER

READER INPUT

Responses to our *Super Cheap Computers* issue, and

More Problems = More Fun

>> My daughter finished up her clock project (Volume 48's "Supersized Seven Segment Clock," page 16) a few weeks ago. She is very proud of herself and had a great time putting it together.

We had a number of difficulties with the project. We got bitten by quite a few bad parts, the signals for one digit were bleeding into the other digits, and the Arduino is inherently a poor time keeper! We solved that problem with a ChronoDot (a time-keeping add-on for Arduino). It was easier than we expected.

Anyway, thank you so much for this great article. If it had all just worked, it would have been far less fun! I mean it!
— *Cliff Lasser, Massachusetts*

Executive Editor Mike Senese Responds:
Cliff, this is great. I love the finished project photo with your daughter! The satisfaction of sticking with a problem, figuring it out, and finally seeing your project work is one of the best parts of making, as I'm sure she's discovered with this.

IN RESPONSE TO MAKE: VOLUME 49

Hello! In Volume 49 of *Make:*, the "Meet C.H.I.P., the $9 computer" article (page 29) has inspired me to use C.H.I.P. in several of my computer projects because of the small cost and great power. Thank you!
—*Jarrod (8th grade), via the web*

I am surprised that you did not mention the Arduino Pro Mini in your table ("Table of Boards" Volume 49, page 111). It is quite powerful, very small, and can be had from Adafruit for $10 but from China for $2–$3 postpaid. I use them by the handful and so far none of the cheap imports has failed to work correctly.
—*James Bryant, via the web*

I just received my (printed) copy of Volume 49 of *Make:* magazine. Great issue! You knocked the ball out of the park! The bacon article is very interesting. Not what I would have expected from a hardware magazine. While trying to be more vegetarian, I admit (lament) that I l-o-v-e bacon! Sigh. How about an article on soap?
—*Bruce de Graaf, Massachusetts*

Production Manager Craig Couden Responds:
Thanks Bruce! Glad you enjoyed the issue. Not only do we have an ever-growing collection of soap projects and articles at makezine.com/go/soap-projects, but you can also check out *Make:* Volume 18's (non-vegetarian) bacon-scented soap recipe (page 139) for the best of both worlds. Find it online at makezine.com/projects/hogwash-bacon-soap.

IN RESPONSE TO "5 FEATURES THAT MAKE A GREAT TECH TOY"

(makezine.com/go/5-tech-toy-features)
Great article! Coming from someone who works for an educational "toy" company and whose brother is a high school math teacher I can certainly relate to the points that [author Kathy Ceceri] makes. We've found through our own research that educational platforms that have a "low floor and a high ceiling" make for an ideal learning environment, *i.e.* there's a low threshold to entry, but a lot of capacity built into the system when students are ready to move on to more complicated things.
—*Caitlin O. Bigelow, via the web*

WHEN YOUNG MAKERS GROW UP

I'm a "young" (24) maker, and for a good chunk of my life I've been loving everything you guys have done. My dad got me a subscription for the first few years and once I have some extra cash I plan on renewing my subscription. I also made a road trip out to Maker Faire in the Bay Area and can't wait to make it back. I've wanted to write you guys for a while to thank you all for what you do. I don't know that I would have started my own shop or got my giant CNC without your inspiration.
—*Isaac Doubek, via the web*

Executive Editor Mike Senese Responds:
Isaac, we're so pleased to have been part of your journey. Please say hi at the next Maker Faire, and be sure to send us photos of your shop and projects: editor@makezine.com.

MAKE AMENDS

In *Make:* Volume 49's "Making Waves" (page 15) the photo was incorrectly credited to Eric Futran. It was shot by David Kindler. Sorry for the mix-up David!

CNC tools, re-imagined.

Introducing the
Handibot® Smart Power Tool
v 2.0 Adventure Edition

This is the innovative power tool you can take anywhere and run from any WiFi-enabled device. Use it to perform precise and easily repeatable CNC operations like cutting, carving, machining, and milling. Work with wood, MDF, plastics, foams, and aluminum. All at the press of a button.

Pick-up-and-carry-portable! Simply place it directly on your material, whether it's on a table, the floor, the wall, or even the ceiling. The tool cuts through its base into whatever you're working on.

Revolutionary new control software, FabMo, unleashes new freedoms. You don't need to work with complex CAD and CAM software; use Handibot apps to make projects and run simple jobs. And you can run the tool from your smartphone, tablet, or laptop. It's wireless CNC!

Grab your Handibot and your smartphone and *go your own route*.

For full tool specs and to purchase, visit www.Handibot.com

carve your path
www.handibot.com

DISCOVER
CREATE
COLLABORATE
INVENT
MAKE with the new Arrow.com

Visit us at Maker Faire Bay Area to learn how we can help you go from prototype to product and beyond.

#MAKEWITHARROW

ARROW
arrow.com

WELCOME

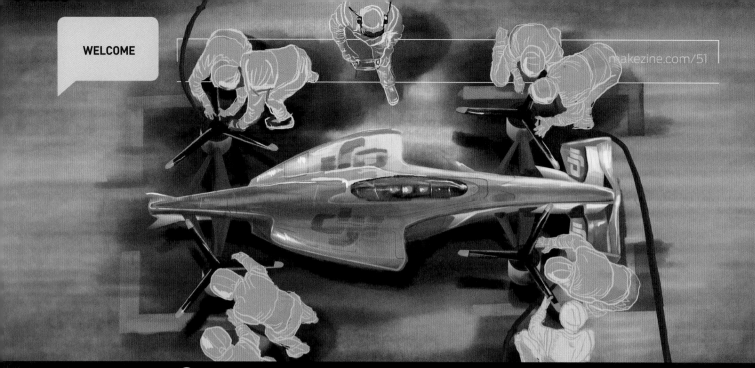

makezine.com/51

Flying into the Future

Makers are raising autonomous aircraft to new heights

BY MIKE SENESE, executive editor of *Make:* magazine

HUMANS HAVE ALWAYS HAD A FASCINATION WITH BUILDING FLYING MACHINES. Ancient Greek lore details the tragedy of Daedalus crafting contoured, intricate wings from feathers and wax for him and his son Icarus. During the Renaissance, Leonardo Da Vinci diagrammed concepts for complex winged contraptions that continue to captivate modern engineers. French brothers Joseph-Michel and Jacques-Etienne Montgolfier invented the hot air balloon in 1783, elevating humans through man-made means for the first time in history. One hundred twenty years later, another pair of brothers, Orville and Wilbur Wright, created the first powered airplane, and the age of aviation was officially born.

In the century since the Wright Brothers, the capabilities of powered flight have advanced with stunning acceleration. Short-distance commercial air travel emerged a few years after their inaugural takeoff. The first nonstop transatlantic flight occurred in 1919, a flight over the South Pole 10 years after that. In 1939 the jet engine became an alternative to propellers; in 1947 we shot past the sound barrier. We landed a man on the moon in 1969, a sophisticated car-sized robot on Mars in 2012, and just this April, a 10-story-tall booster rocket successfully landed itself — upright — on the deck of a self-navigating barge at sea. From these capabilities, massive industries have been born — transportation, defense, research, and to a smaller degree, entertainment.

We're now starting down a similar path with Unmanned Aerial Vehicles. The military has used large, self-piloting aircraft for a quarter century, but small, personal drones in versatile multicopter configurations have only been accessible for a few years (see page 18). The drone community is now determining the best ways to utilize these craft, and while the initiatives mirror those of manned aircraft, the current areas of development are somewhat different.

Right now entertainment is the most mainstream — and DIY — use for drones. Smaller, cheaper, and safer than helicopters, they're useful for cinematography and photography. But their speed and agility make them perfect for racing, and exciting FPV racing videos continue to inspire new pilots to build their own machines. Over the past year, we've seen the first national drone racing event, held in Sacramento last July, and a World Drone Prix race this March in Dubai, where 15-year-old Luke Bannister won the $250,000 prize purse. Our popular drone section at Maker Faire has grown as well; this May at Maker Faire Bay Area our friends from the Aerial Sports League will be hosting a 16,000 square-foot flying zone.

Like early aviation, development of UAVs and their applications will progress at an incredible rate. Cargo delivery (transportation) and industrial inspection (research) are being pioneered by individuals, but have huge interest from powerful companies. These promise to grow into substantial categories once they overcome obstacles like speed, distance, accuracy, and regulation. Personal transportation is a possibility; Shenzhen-based Ehang even debuted a person-carrying autonomous quadcopter at CES this January, although it will be years for the FAA and general public to be comfortable with the concept.

As for racing, hackers and hot-rodders will continue building new rigs to out-gun each other through the lenses of their onboard cameras. Organizers will need to pull more spectators to keep the prize money big, though, and to do that, the leagues will need to raise the stakes. Larger drones and faster speeds will establish new categories and allow for more viewer-friendly races with the edge-of-the-seat exhilaration felt in today's multi-billion-dollar racing industry. They'll need larger courses, well-heeled sponsors, and teams of engineers and pilots.

All this comes back to enthusiasts in their workshops, Makers like you cranking away on propellers and brushless motors. Our fascination with building flying machines has no limits. Our predecessors built modern aviation; you are creating its future.

James Burke

ENGRAVE IT. CUT IT. MARK IT.
The only limit is your imagination.

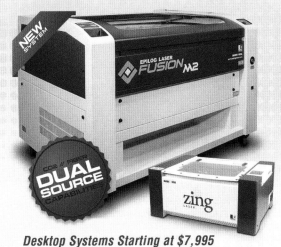

Your vision combined with Epilog's power equals spectacular creations.

From laser die cutting, to engraving photos, to marking parts and gadgets, our laser systems create the products you see here and more! Whether you're working with wood, acrylic, paperboard or just about anything else, Epilog Laser systems precisely cut and etch the components you need.

Desktop Systems Starting at $7,995

Contact Epilog Laser today for a laser system demo!
epiloglaser.com/make • sales@epiloglaser.com • 888-437-4564

MADEiNUSA
Golden, Colorado

Infinitely Hackable
Practically Indestructible
Available in the Maker Shed

MADE ON EARTH

Backyard builds from around the globe
Know a project that would be perfect for Made on Earth?
Email us: *editor@makezine.com*

BEHANCE.NET/LISA_SMIRNOVA

PAINTING WITH THREAD

While her full-time job consists of commissioned illustrations, interior paintings, and conducting weekend classes, Moscow-based artist **Lisa Smirnova**'s favorite "painting" medium is embroidery. Her post-impressionistic stitching is reminiscent of Van Gogh, but instead of chunky layered paint, her work draws its gorgeous texture from the interwoven threads of richly colored yarn.

"I like working with clothes, textures of fabric, and mixes of colors," she says. "My work is more emotional than deeply meaningful. I've never tried to raise any problems — I try to make it in a way that satisfies me emotionally."

Smirnova sketches some ideas on paper and chooses a favorite to scale appropriately as a template for embroidery. She transfers the image to fabric using a pencil, and sews the outline before filling it in. As with watercolor, Smirnova chooses a color palette and adds hues from dark to light.

It takes her anywhere from three days to three months to finish her embroideries, which have found homes on everything from high fashion to pillowcases and pullover sweaters.

— *Sophia Smith*

CYBERNETIC CERAMICS

BRENDANTANG.COM

In the nearly 80 pieces of **Brendan Tang**'s sculpture series *Manga Ormolu*, hunks of futuristic robots live among the curved lines and intricate patterns of Chinese ceramics. Tang's juxtaposition is inspired by ormolu, the 18th-century European practice of adding gold gilt to existing art pieces. But the charm of his series lies in the difficulty of the two trying to mesh, as robotic elements routinely squish the elegance of the pottery, and the pottery overflows the robotic constraints.

"When I originally saw an example of ormolu, I was drawn to the act of cultural appropriation and hybridization. In their efforts to create a curiosity object I saw my own story of an immigrant adapting to Western culture," Tang said. "I also wanted to mash up my two loves: ceramics and giant robots."

Each piece is made from low-fire white ceramic, metal, wire, glass, and plastic. He uses traditional ceramics tools like potters wheels, slab rollers, and clay extruders to get the general form, and then uses metal ribs and wooden modeling tools for finer details. Tang hand paints the blue and white patterns and airbrushes the robotic elements.

A single piece takes one to three months, though Tang usually works on more than one at a time. At their largest, the pieces can be up to 30 inches tall and 40 to 50 pounds.

— *Craig Couden*

makezine.com 13

MADE ON EARTH

BURNING DESIRE SANDMANCREATIONS.COM

In 2001 **Sean Sobczak** left Burning Man feeling inspired. He had never built a thing. "I bought a few hundred feet of wire and a few tools, went home, and began bending the wire. I had no idea what I was doing, but I was learning every day and really enjoying the process." His first projects were three illuminated sea horses. From there the sculptures kept getting larger until he built his 25-foot-long dragon, which has 700 "scales" and is lit with 12,000 Christmas lights.

When embarking on a project of this magnitude, Sobczak first sketches out a basic outline and writes lists of details he wants to include. He welds together the internal structure with support for armature and lighting, and makes sure that there's a way to break it down for transportation. The details are accomplished by working the wire with his thumbs and a pair of pliers. For a final touch, Sobczak adds lights and fabric. "This softens everything up by adding some color, and the wire work, backlit by the internal illumination, shows up as dark lines through the fabric skin."

— Lisa Martin

MODERN MARIONETTES
PUPPETSINPRAGUE.EU

Puppets may seem like child's play, but you can't underestimate the artistry and engineering that go into creating a classic marionette. **Mirek Trejtnar** could tell you: He's carefully researched their anatomy, borrowing specialty puppets like skeletons, acrobats, dragons, and more — some over a century old — from museums and private collections. In order to get the proper balance between complexity and mobility, he studies methods of stringing the puppets and arranging their systems of joints.

Trained as a woodworker at the top fine arts high school in communist Czechoslovakia, Trejtnar spent a short stint restoring furniture in Baroque churches, but he soon chiseled out a life as an animator, artist, and now a modern-day Geppetto in the heart of the kingdom of puppets. But unlike Pinocchio's dad, Trejtnar and his American-born wife Leah Gaffen share detailed instructions for puppet making on their website. Their online readers often come to Prague for popular hands-on workshops. These crash-course apprenticeships end with delightful performances by an international troupe of newbie puppeteers. Gaffen says, "There's nothing like seeing master carvers work — it's the best way to learn. The most magic moment is when all the pieces of the carved puppets are put together and the puppet is suddenly alive."
— *Michelle Hlubinka*

FEATURES

Maker ProFile Written by DC Denison

Crowdfunding Coach

Helpful advice from Kickstarter's John Dimatos

John Dimatos is the senior director of design and technology communities at the crowdfunding service Kickstarter. He works with project creators — aka Makers — in the product design and tech categories, providing feedback and answering questions. Before he arrived at Kickstarter in 2013, he was head of applications at MakerBot Industries. John offers the following advice to Maker Pros.

BEFORE YOU LAUNCH A KICKSTARTER CAMPAIGN …

Prepare. Be really clear about why you are doing it and what you expect to get out of it. When launching a project, creators often focus on getting money to buy parts to get things done. But it's good to remember that the money represents people. Will you have 10 backers, 100, 1,000? Each number requires a different kind of communication. Emailing and talking to 1,000 people is very different from talking with 10 people.

HAVE A PLAN FOR AFTER THE CAMPAIGN

This matters a lot. For instance, if there's manufacturing involved that requires third parties, know who that will be, have a lot of conversations with them, understand their capabilities, what expectations they have from you, and how they treat their employees. Not knowing these things before you launch the campaign puts you in a tough position afterward, when you're making decisions on the spot.

WHY KICKSTARTER IS "NOT A STORE"

When we say "we're not a store," we mean that you don't come here just to make a monetary transaction. You believe in an idea, in the person who's making it, the larger ecosystem that supports it. Kickstarter is a reminder that what we do with our time, our money, our lives — that should mean more than just buying a microwave oven in a big-box store. People are backing projects to support an idea. There's no guarantee that projects are going to work out. Supporting something that doesn't exist yet is part of the mantra here.

HELPFUL DATA FOR MAKER PROS

We have a great data team that looks at the numbers and helps us advise our creators. For instance, a typical campaign has 5 to 7 tiers for backers — we know that's an easily digestible number. The most typical pledge level is $20–$25. Within the tiers, the $100 tier tends to provide the most funds to a campaign. If a campaign hits 20% of its goal, it has an 80% chance of success.

ASK YOURSELF: DO YOU WANT TO COMPLETE A PERSONAL PROJECT OR START A BUSINESS?

If you're aiming to one day make your Kickstarter project the main thing you do, you have to plan for that. If you're assembling 1,000 units at your house, it will get done, but you're giving up your ability to focus on how to build the business you're trying to make.

On the other hand, maybe you don't want to build your project into a bigger business. Maybe you just want to sit around making things with your friends.

We encourage people to think about which Maker they want to be. We love both, and both are necessary for a healthy ecosystem.

DO YOU EVER THINK ABOUT LAUNCHING A KICKSTARTER CAMPAIGN?

I think about that every day! It would definitely involve LEDs. I love lighting projects.

DC DENISON is the editor of the *Maker Pro Newsletter*, which covers the intersection of Makers and business. He is the former technology editor of *The Boston Globe*.

For more Maker Pro news and interviews, visit makezine.com/go/maker-pro.

2016

#Tormach2016

Announcing the 3rd Annual

TORMACH OPEN HOUSE

When: Saturday, July 23, 2016
Where: Waunakee, WI
Agenda: www.tormach.com/openhouse
Cost: FREE

CNC Machining Seminars

Product Demos

Tormach Technical Staff Q&A

Guest Speakers

Customer Showcases

Meet & Greet Reception

Come check out the new PCNC 440!

RSVP Now at:
Tormach.com/openhouse

Over a million ways to give your business a major competitive advantage!

SuperSpeed

Competition

Gain a competitive advantage with SuperSpeed™ – designed to dramatically improve productivity.

Leverage our 27+ years of continuous innovation with market-first technology only available from Universal Laser Systems:

- Complete laser material processing ecosystem by design
- Modular architecture
- Rapid Reconfiguration™
- Dual laser platforms
- Air-cooled lasers from 10W to 500W
- **SuperSpeed™**
- MultiWave Hybrid™ technology
- Class 1 to Class 4 transformable platforms
- Integrated advanced air filtration systems
- Fire suppression
- Laser Materials Processing Database
- Advanced Process Control Software
- And more...

UNIVERSAL LASER SYSTEMS®

To find out how to create your own unique laser cutting, engraving and marking system from over a million available combinations, contact us today!

ulsinc.com/superspeed_108
1.800.859.7033

©2016 Universal Laser Systems. All rights reserved. Universal Laser Systems' name and logo are registered trademarks of Universal Laser Systems, Inc.

Special Section

DRONE R-EVOLUTION

Written by Mike Senese ◆ Illustrated by Rob Nance

THE DRONE FIELD HAS CHANGED DRASTICALLY. Just a few years ago, the only way to get a multicopter was to build your own, relying on community guidance from sites and resources like DIY Drones and the open-source ArduPilot development.

The 2010 launch of the Parrot AR.Drone brought pre-built quadcopters to the mass market. Since then, there has been an arms race between manufacturers. Every release includes new, advanced features and powerful processing that makes these machines so easy to fly that they now fly themselves. They perform almost any type of aerial maneuver you'd want, at the push of a button. The latest of these, DJI's Phantom 4, can even steer around obstacles in its way. With their rock-solid stability and high-quality cameras, pilots are now using these for countless purposes. The best part? We are just getting started.

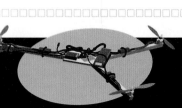
WOODEN TRICOPTER — 2007

BASIC QUADCOPTER — 2007

PARROT AR.DRONE — 2010

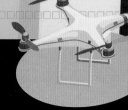
DJI PHANTOM — 2013

Special Section
DRONES ◆ SEARCH AND RESCUE

Written by Patty Hodapp

AERIAL

Matthew Billington

makezine.com/drone-revolution

AID

PATTY HODAPP is a journalist reporting on fitness, adventure, and more. She writes and edits for several national magazines.

AMATEUR DRONE PILOTS CAN SAVE LIVES, FIND BODIES — OR WREAK LOGISTICAL AND LEGAL HAVOC

JIM BOWERS, DRONE HOBBYIST AND ARTIST, GOT A CALL LATE IN THE AFTERNOON ON A SNOWY WINTER DAY. Tiffany Matthews, an acquaintance from their Colfax, California, community was on the other end, desperate for help. She needed his drone to find her missing fiancé, and quick.

"I could hear panic in her voice," says Bowers. "The family was looking at all options, so I didn't think about it for more than 30 seconds before I decided I needed to at least try."

Eric Garcia, 39 and a father of two, disappeared on Saturday, Dec. 7, 2013, somewhere between El Dorado and Placer Counties. He left Rancho Murieta in his tan '99 Plymouth Breeze, braving the icy roads and six inches of snow to retrieve his wallet from his home in Colfax, 50 miles away. He never arrived.

A search and rescue (SAR) mission involving the two counties' sheriffs' departments ensued, and was subsequently suspended after they found no trace of Garcia. Matthews, frustrated, called Bowers. "I don't know why she thought about using a drone, but she knew I was the only one flying in Colfax at the time," says Bowers. "At first, I was baffled [Matthews called] because I'm a lifelong obsessed artist. I use drones to create video documentaries. So when she asked me if I would do something so serious, it was a little bit of a shock."

Using his DJI Phantom 2 Vision+, Bowers manually covered a 40-mile area for a week, in 4- to 5-mile sections, targeting cliffs, embankments, and places volunteers couldn't reach on foot. The Garcia family agreed to place Starbucks coffee cups on the sides of roads marking unreachable areas Bowers should search with the drone.

He narrowed the area to a three-mile spread between Weimar and Colfax, upon which the sheriffs' departments resumed the search for one more day. There they found Garcia in his wrecked car down a steep dirt bank off the side of Interstate 80. He had hit a tree and died on impact.

"The outcome ... in the end just kind of freaked me out," says Bowers. "Everybody was very, very grateful for what I was able to do, and it gave the family hope and closure." But to Bowers, Garcia was just the beginning. "It moved me so much that I came up with idea of SWARM [Search With Aerial RC Multi Rotor] where volunteer pilots could register and I could dispatch them to search, helping the families of missing persons."

LAWS AND LOGISTICS

SWARM now has more than 3,000 drone pilots registered worldwide, and at least one in every U.S. state. Its Facebook group alone has over 5,100 members.

But while SWARM's goal to help families find their missing persons is well intentioned, there's little his pilots can do legally to help until after the official SAR investigation has been terminated

HOW TO VOLUNTEER YOUR DRONE FOR SEARCH AND RESCUE MISSIONS

As Federal Aviation Administration guidelines stand, do-good hobby pilots cannot legally fly their drones to assist official search and rescue workers. While this will likely be addressed by the FAA as they write another round of UAS-law, and public and governmental agencies are currently using drones, a community of civil pilots is already poised to use their aircraft to help.

If you want to get involved, remember that you must wait until an official investigation is complete. But even then, don't just run out into the field with your drone. This stuff is complicated and the law leaves room for interpretation, which can result in a mess. *Make:* tapped a mix of civil and public pilots for general search and rescue rules to know before you go out on your own.

BEFORE YOU GO: RESOURCES AND PREP FOR CIVIL SEARCH AND RESCUE PILOTS

First, get educated. Join a forum, like SWARM's Facebook group, for discussions about equipment, programs, and search techniques. In addition, study up on best practices for data collection like those published by CRASAR. And get trained in the workings of the National Incident Management System, as it will likely manage the operation. Knowing how the system functions will help you ascertain where officials had already searched before ending the official operation, which will save you time.

(Continues on page 23)

Raul Arias

Special Section
DRONES ◆ SEARCH AND RESCUE

by the Incident Command System — the government agencies (sheriff, police, fire) in charge.

There's no question that drones have become a proven tool in the SAR field. They can collect high quality data doing logistical legwork for rescue workers, and in recent cases, they have actually saved lives. In May 2013, using a Draganflyer X4-ES drone with thermal imaging, the Royal Canadian Mounted Police in Saskatchewan saved a man whose car had flipped. In July of 2015, rescue officials used a drone to deliver a life jacket to two boys stranded on a rock in the middle of the Little Androscoggin River, near Mechanic Falls, Maine.

But it's not as simple as running into the woods with a drone and expecting to contribute. Collecting and processing data is complex and it's easy to miss important details. Typically, pilots use a program like DroidPlanner to automate a grid pattern, often miles wide. Using mounted cameras or thermal imaging sensors like those from FLIR, and FPV (first person view) technology, they record video, which can be reviewed later at 2x speed. They fly 60–100 feet off the ground (any higher, and the missing person becomes too small to spot) and shoot in 4K resolution. Flights last 10-25 minutes, giving the pilot enough time to scan the area in zigzag-sweep or expanding-spiral search patterns. SWARM allows crowdsourcing, where pilots can upload footage to the Facebook forum and ask for multiple sets of eyes to comb through it.

This yields several problems: Uploading search footage may mean exposing

> There's no question that drones have become a proven tool in the SAR field.

personal, sensitive, or even graphic images. And no one knows how best to meld public (government) and civil (non-government or hobbyist) pilots into the SAR process. What's more, key stakeholders, like SWARM and public officials, are not talking at the national policy level. They aren't speaking the same language. Yet.

Some, like Dr. Robin Murphy, professor of computer science engineering and director of the Center for Robot-Assisted Search and Rescue (CRASAR) at Texas A&M University, think SAR missions should be left to the professionals. Flights by civil pilots, she says, "just might not be the best use of hobbyist time in helping with SAR right now. We say in Texas, you can have a gun and you can carry a permit, but you don't get to go to a police shootout and help, because you weren't deputized. I think it's the same thing with drone hobbyists right now."

Murphy's program, Roboticists Without Borders, offers free drone-forward disaster assistance to public SAR efforts. "Very few missions are going to be like the use of the Draganflyer to find the guy in Canada," she says. "I've been doing SAR work and thinking in the field since 1999, and a lot of the things that you use drones for in SAR are not for direct life saving because they don't replace people and dogs." Mostly, drones are used to look where volunteers can't reach, and rule out where missing persons are not — as Bowers did in his search for Garcia.

SETTING STANDARDS

But if civil pilots are to work in tandem legally with public operations without compromising the official mission, a nationally accepted set of best practices is a must. As the barely-a-decade-old industry booms, drone use in the SAR field has soared as well. The challenge for the Federal Aviation Administration will be legislating fast enough to define the role of drones in the industry. Until then, it's limited

1

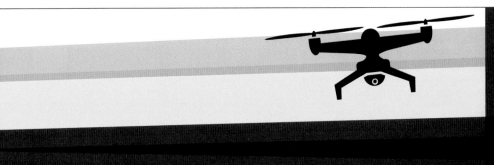

to a conversation between hobbyists and officials, and that conversation isn't exactly flourishing. "I'm not trying to make people mad and I think there is a role for general [civil] pilots, but I'm not exactly sure what that is, yet," says Murphy. "We have to work through what that looks like, and the responsible people will step up to the plate, and help figure it out."

The FAA set its guidelines for small, unmanned aircraft systems late last year. The provisions address two main safety issues: Keeping unmanned aircraft systems (UAS) clear of manned aircraft, and mitigating risk to people and property on the ground. They clarify the exemptions to obtaining a certificate of authorization (COA) to fly, and officially define what constitutes a small UAS, who can operate one, and under which circumstances.

The public SAR sector, meanwhile, has already moved to the next phase: government agencies are adding drones to their toolkits. Some are already actively using drones. In August 2014, the Austin, Texas, city council approved a four-year drone study to establish best practices in fire and disaster situations with the Austin Fire Department. Many other operations have COAs to fly or have an application in to the FAA to get one, but the process can take up to 90 days — clearly a problem when conducting emergency searches. As a workaround to the delay, the FAA launched a new policy in March last year which grants COAs to commercial operations, like public SAR groups, with a Section 333 exemption (this takes 120 days, but then allows the FAA to distribute COAs quicker).

"Almost every SAR organization I know of, have heard of, or come across is either using drones or are actively pursuing a 333 exemption to fly," says Gene Robinson, former chief drone pilot for the National Institute of Standards and Technology and founder of RP Flight Systems, which has developed unmanned aircraft for rescue and law enforcement agencies since 2001.

"We certainly deal with a society who does not want to be regulated because they think search and rescue is easy, and I can assure you, this is not easy," says Jerry Hendrix, chief engineer and executive director of the Lone Star UAS Center at Texas A&M University of Corpus Christi, an FAA test site designed in February 2013 to integrate unmanned aircraft into the national airspace. "It's important to know that there are rules that we follow for the protection of all. The number one goal is safety and it's crucial people learn to work within the system to be able to support SAR efforts with UAS."

Bill Quistorf, chief pilot for the Snohomish County Sheriff's Office in Washington, has worked with state, military, and federal aviation agencies for 45 years, as well as Murphy's CRASAR. He also ran high altitude SAR operations in Alaska, and says communication is the biggest issue facing the industry. "I'm pro drone but I'm also realistic. People always want to tell me how wonderful drones are, and I say, 'Yeah, yeah, I understand, but look at the limitations, don't just think of all the things they can do'," says Quistorf. "They can augment the SAR helicopters but they can't replace them — and if we're going to augment the helis

GET THE GEAR YOU NEED: HOW TO TRICK OUT YOUR COPTER FOR SEARCH AND RESCUE

Whether it's fixed wing or a quadcopter, you may want to equip your drone with a first-person-video rig, or opt for a recording to view later. Opinions differ, but either way, you want to take a systematic approach using grid-flying software like Visionair or another flight control system, and use a camera that shoots high-resolution footage. Map photos at regular intervals. Study the terrain, and keep a rigorous log of anything abnormal. An infrared camera can be another useful addition. Software such as Pix4D or AgiSoft PhotoScan can use photos snapped at regular intervals, combined with latitude, longitude, altitude, heading, bearing, and more, to produce a mosaic image of the entire area covered.

In addition to selecting your equipment, you'll have to prepare for weather, terrain, and being outside in remote areas for hours. Merle Braley, a hobby pilot and creator of SWARM's website, has a backpack ready with essentials: 12 charged batteries for the transmitter, a notebook to record battery life and track the time, FPV goggles, a 4'×6' army green blanket to spread out his gear on, a fluorescent orange vest to wear when he flies, a poncho to cover the gear if it rains or if the ground is wet, waterproof hiking boots, and energy bars. He also brings binoculars and a spotter — a buddy who can keep an eye on the drone and communicate interruptions so the pilot can remain focused.

1. SWARM founder Jim Bowers geared up for a mission with quadcopter controller and FPV headset.

Special Section — DRONES • SEARCH AND RESCUE

SEARCH AND RESCUE DOS AND DON'TS

- **DO: Plan smart, and expect the unexpected.** Often the demand for search and rescue hobbyists is a last resort, and urgency is of the utmost. Have your gear organized to leave at a moment's notice and don't deploy haphazardly before you consider weather conditions — for the drone and for yourself — and how to set the grid pattern.
- **DO: Be prepared, physically and emotionally.** Rescuers shouldn't have to rescue you.
- **DO: Practice.** SWARM has local groups that practice in challenging areas, hiding dummies and searching them out with drones.
- **DO: Know the rules.** Know your limits, where and when you can fly, and don't push them. If you're not charging for your services, then you just have to follow the FAA's rules regarding model aircraft use. Don't fly more than 400 feet above the ground (missions should take place between 60-100 feet anyway), keep the drone within sight, and give airports a five-mile radius.
- **DO: Check with the landowner** if you need to enter — or fly over — private property.
- **DO: Prioritize safety.** That means your safety, the family of the missing person's safety, and the safety of anyone on the ground in your search area.

- **DON'T: Go in cold, or deploy in doubt.** Not just anyone can throw a copter up in the air and expect success. Do your homework with your local fire and sheriff's departments.
- **DON'T: Underestimate the power of your data.** While pilots don't always agree what type of data is best, it's often the processing of that data that leads to discovery. No matter what you collect, it's crucial the information is catalogued in clear, digestible chunks. With hundreds of images or video you'll want to separate them into sections of land — create a folder on your computer and break that into sub folders, one for each easily identifiable area. Label the images to isolate latitude and longitude on a picture more quickly, and so you can put a searcher in close range of the target in the picture. Here, your terrain log will come in handy to provide written detail when you review the images.
- **DON'T: Be the hero during an official investigation.** You must wait for the SAR government operations agencies to pull out. Period. Unless you have a relationship with officials in your area, a Section 333 exemption and COA, previous flying experience, and you know how the incident command system (IC) works, steer clear. Getting in the way could result in a felony. But once the official investigation has ended, you get the green light, legally.

1

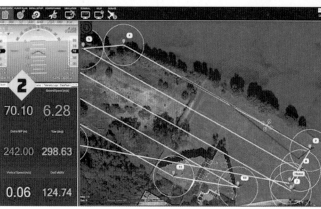

2

3

we need a plan in place and we have to do it safely."

Most of Quistorf's SAR missions take place in the mountains, dozens of miles from any road, logistically posing problems for potential drone involvement. "You can't fly a hobbyist drone 30 miles from the road and keep track of it," he says.

INCREASED DEMAND

Despite apprehension from Murphy and other SAR officials, the civil SAR pilot sector is growing rapidly. SWARM adds around 50–100 new members per week to its private Facebook forum alone. As its membership and reputation go up, so has the demand for its involvement from families of missing persons. "In the beginning, we got calls once a month, and now we get requests at least twice a week," says Bowers.

Little did he know, when he dispatched himself to search for Garcia, Bowers started a worldwide movement in search and rescue. And though SWARM's operations are legal, its very existence poses a larger question the FAA and all public SAR agencies must answer — what is the role of drones in SAR, and how, if at all, can hobbyist pilots help?

One step is for SWARM to enforce best practices of its own to maintain the organization's reputation and promote the technology, despite blurry FAA guidelines and a drone-shy culture. Bowers wrote standard operating procedures for SWARM, a code of ethics that follows FAA law, and integrated an assessment interview to determine pilots' skills, equipment, personalities, professionalism, and commitment before he dispatches them to the scene.

"Right now anyone can buy a drone," says Bowers. "Any idiot can put them up in the air, and some will do it over an airport or wildfire and will screw it up for the rest of us." To counter negligence, SWARM encourages its members to reach out to their local fire and police departments to build relationships with officials. To keep their skills fresh, localized groups regularly practice rescue missions using human dummies dressed in earth-tone clothing. Bowers frequently attends expos to keep updated on the latest drone technologies.

In addition to search and rescue, Bowers films and produces videos and documentaries using his fleet of 18 drones. In a small security shed on his California property, he edits and publishes material on his drones-for-beginners YouTube channel, Demunseed (he's got 24,068 subscribers and 2.4 million views). He happened to be the last drone pilot to legally shoot a documentary in Yosemite with his DJI Phantom, titled *A Drones-Eye View*, before the FAA banned drones in national parks in 2014. He has also built 30 UAS, for himself and others, in this workshop. He became obsessed with the search for better cameras, and the newest gimbals to mount them.

"If drones didn't have cameras on them, I would have lost interest," says Bowers, who flew R/C airplanes for two decades before drones became available. "But the technology came along, and I started building my own from scratch. When I figured out that you could put a camera on them and see the world from the top down, I was hooked."

He also creates big, public art. Among large-scale murals and sculptures, he

> "In the beginning, we got calls once a month, and now we get requests at least twice a week"

1. Dr. Robin Murphy and students conducting practice flights.

2. DroidPlanner and other software options let pilots create autonomous routes to search an area without any hands-on control of the aircraft.

3. A DJI Inspire 1 quadcopter scans past a remote road.

4. Bowers and a SWARM crew launch their SAR drones.

makes statement pieces for festivals including Burning Man and Coachella. He holds the Guinness World Record for building the largest working timepiece on Earth — a 1.25-mile-diameter clock he constructed at Burning Man in 2012.

Moved by his experience with his search for Garcia, Bowers planted 55,000 daffodils on Colfax's westbound off ramp to I-80, using profits from an annual festival he spearheads for the community. The perennial flowers bloom each spring and stand as a memorial to Garcia. "It's coming up on the three-year anniversary of his death, and it's always a hard time for the family," says Bowers. "But the flowers are beautiful. My way to give back, maybe. It's just a huge carpet of yellow alongside the highway."

Learn more about SWARM at sardrones.org

Special Section
DRONES ◆ BOOK EXCERPT — GETTING STARTED WITH DRONES

LITTLE DIPPER DRONE

Written by Terry Kilby and Belinda Kilby

GAIN VALUABLE KNOWLEDGE OF AERIAL ROBOTICS AND DEVELOP SOLDERING SKILLS WITH THIS 300-CLASS AUTONOMOUS QUADCOPTER BUILD

TERRY AND BELINDA KILBY
are drone enthusiasts, aerial photographers, Makers, trainers, and a husband-wife team. They have been designing and building small unmanned aerial vehicles (UAVs) for artistic and practical aerial photography since 2010 through their company, Elevated Element.

MATERIALS
» **Little Dipper Airframe** CNC files available at makezine.com/go/little-dipper-build; most pre-made FPV frames will also work.
» Power distribution board (PDB)
» Electronic speed controllers (ESCs) (4)
» Brushless motors (4)
» Bullet connectors (optional), 2mm, male and female pairs (12)
» Heat-shrink tubing, ⅛"

TOOLS
» Soldering iron and solder
» Helping hands or other clamping system
» Heat gun or hair dryer
» Wire cutters/strippers
» Needlenose pliers
» Allen wrenches
» Small zip ties
» Double-sided foam tape
» Hobby knife and scissors
» Fine-tip marker or paint pen

IN OUR NEW BOOK *MAKE: GETTING STARTED WITH DRONES*, WE GUIDE THE READER TO ASSEMBLE THEIR OWN FPV-STYLE QUADCOPTER FROM SCRATCH. In this excerpt we start the first-time drone builder on the construction of the frame and electronics for the Little Dipper, a 300-class autonomous flying rig.

The compact Little Dipper's airframe is made up of two subframes that help isolate the motor vibrations from the flight and imaging sensors. These subframes are called the *clean* and *dirty* frames. The dirty frame is the bottom subframe, and it holds all the moving parts, such as the motors and the propellers. The clean frame sits on top and holds all the flight and communication electronics. The folding arms soften the impact in the event of a crash.

1. BUILD THE LITTLE DIPPER AIRFRAME

The frame is open source; you can download the laser- or CNC-cuttable design files from makezine.com/go/little-dipper-build, where you can also find step-by-step video instructions. The following steps also work with most commonly available FPV racing drone frames.

2. MOUNT THE POWER DISTRIBUTION BOARD

The PDB lets a single battery send electricity to all the drone's electronic components — speed controllers, motors, camera, etc. We made our own PDB from copper-clad G-10 (Figure **A**), but there are many small, inexpensive versions on the market.

Start by applying a couple of small strips of double-sided tape to the back of the PDB.

Next press the PDB into place in the middle of the dirty frame where it will be easily accessible by the battery lead, electronic speed controllers (ESCs), and any accessories that need access to power (Figure **B**). There's a 3mm hole in the middle if you'd like to add a screw for extra support. We found that the double-sided tape did a very good job and opted to not use a screw. If you do use a screw, try a small nylon screw and nut — it will save on weight and and it won't act as a conductor.

3. SOLDER ON THE BULLET CONNECTORS

This step is optional, but it can make the install a lot easier. Bullet connectors allow you to plug and unplug the ESCs and motors into each other rather than soldering them directly. The pros to using them include ease of use during maintenance, troubleshooting, and upgrades. The cons include failure due to loss of contact. If a bullet connector fails, it can cause a crash (one motor out of four stops spinning and you fall like a rock). With this list of pros and cons, you can understand why people have sharp opinions about these connectors. We'll let you decide if you want to use them, but this project will assume that they are installed. If you decide not to, we recommend that you directly solder your connections and seal them with heat-shrink tubing. Just make sure they're right before you fire up that iron!

Bullet connectors, like almost every other type of connector, consist of a pair: one female and one male. You'll be installing the male ends on your motors and the female versions on the ESCs. This is considered a best practice, as the ESC is the end providing the power and the female bullet will be shielded to provide protection when things are not plugged in.

Begin by taking one of your 4 motors and stripping away about ⅛" of the insulation from each of the 3 motor leads. Next, tin the wire tips by adding just a bit of solder to the tip of your iron and coating the outside of your motor leads with it (see Figure **C**).

Next, use your helping hands tool to solder the male bullet connectors to the motor leads. Clamp one bullet connector into one of the alligator clips and one of the tinned motor leads in the other. Once you have everything configured, place your iron on the outside of the bullet connector, allowing it to heat up for just a few seconds before applying some solder to the inside of the connector where the wire sits (Figure **D**).

Once the solder has cooled, remove the motor lead and bullet from the helping

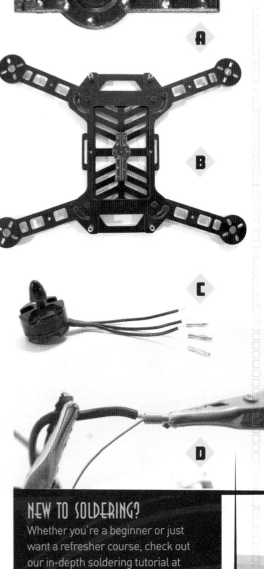

NEW TO SOLDERING?
Whether you're a beginner or just want a refresher course, check out our in-depth soldering tutorial at makezine.com/go/soldering-tutorial.

hands and solder the other 2 motor leads and bullet connectors the same way.

Repeat these steps for the remaining 3 motors. When you're finished, you should have 4 motors with 12 male bullet connectors soldered to all the motor leads (1 on each lead).

Now it's time to insulate your soldered connections. Cut three ½" sections of ⅛" heat-shrink tubing and loosely fit them over, but without blocking, your newly soldered bullet connections (Figure **E** on the following page). Carefully trim away any obstruction.

Special Section
DRONES ◆ BOOK EXCERPT — GETTING STARTED WITH DRONES

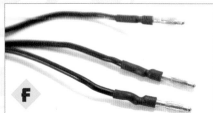

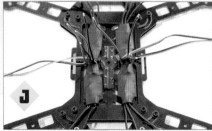

Once you have the heat-shrink in position, apply heat with your gun to shrink it. Apply this process to all 3 leads and you should end up with something that looks like Figure **F**.

Now it's time to do it all over again with the female connectors on the ESCs. Depending on what ESCs you're using, they may already have bullet connectors soldered on. If yours do have connectors in place already, check that they work with your male motor bullets. If everything seems to connect nice and snug, you can skip the rest of this step.

As with your motors, find the 3 black leads on your ESCs (the raw wires, not the servo plug), strip about 1/8" of insulation, and prepare the wire for soldering. Follow the exact same soldering steps you did for the motors, but with the female connectors (Figure **G**).

After all the connectors are soldered, cut 1" or more heat-shrink to insulate each. The heat-shrink should go just to the tip of the connector while still extending over the wire on the other end, allowing the male connector to make a solid connection.

At this point, you have 4 motors with male bullet connectors and 4 ESCs with female bullet connectors. If you haven't done so already, try plugging them into each other and see how they fit (Figure **H**).

4. MOUNT THE SPEED CONTROLLERS
Electronic speed controllers are typically mounted in one of 2 ways: on the frame itself, or on the booms near the spinning propellers to get additional cooling from the downdraft. Because the booms fold on this particular frame, you'll be mounting the speed controllers on the inside of the dirty frame. Apply a small strip of double-sided tape about 1/2" wide to a single side of each of the 4 ESCs (Figure **I**). Sometimes ESCs can have a large round capacitor that sticks up on one side. If that's the case with yours, apply the tape to the other side to get as much surface coverage as possible.

Next, position the ESCs in the subframe. Take one of the ESCs and make sure the tape is peeled back and ready for mounting. Locate the open space in the dirty frame around your PDB. Mentally separate this into quadrants and place each ESC into its own space. The red and black power leads coming off the ESC should be pointing toward the center of the frame, while the black motor leads you soldered in the last step point outward (Figure **J**).

5. SOLDER UP THE POWER SUPPLY
The concept here is to connect the positive and negative leads (red and black wires, respectively) from each of the ESCs in a parallel circuit. If you aren't familiar with a parallel circuit, that's OK. It just means that all the red wires (positive) are joined together in one connection while all the black wires (ground) are on another connection. Figures **K**, **L**, and **M** show how that works: There's one strip on the board for positive leads and one for ground. All of the ESCs as well as the main battery lead will connect to the PDB.

Start with a single ESC. Take the red wire coming out of your ESC and determine how long it needs to be in order to effectively reach a positive circuit on the power supply (in our case, the left-hand strip). Clip that wire to that length (or just a tiny bit longer, just in case) and strip off 1/8" of insulation from the tip. Now tin the exposed wire with your soldering iron and get it ready to be attached to the PDB.

Next, take your needlenose pliers and use them to hold your ESC lead onto the PDB at the point where you want to make the connection. Make sure you are on the correct PDB circuit. This is our positive lead, so make sure it's on the positive circuit. Finally, apply your hot iron (be sure to keep the tip clean!) to the top of the positive lead, sandwiching it between your iron tip and the PDB. If you have applied enough solder to all the components, they should all melt together with no problem. Once that

FUTURE-PROOFING YOUR DRONE
It's good to plan ahead by adding an extra, unused power lead to the PDB for items you want to include down the road. In Figure **M**, you'll notice that we have done exactly that. Try adding a JST power lead (readily available online for a few cents) to your power circuit and leaving it tucked neatly between the clean and dirty frames. Then when you want to add something like a video transmitter, all you have to do is pull that plug out and tap into your power right there. No need to get the soldering iron out again!

happens, remove your iron while continuing to hold the lead for a few more seconds with the pliers. If you pay attention to the solder, you'll see it cool in a matter of seconds. It will take on more of a matte finish look and less of a liquid appearance. Once this has happened, you can remove the pliers and check the connection. If it appears to be loose at all, repeat the necessary steps until you have a solid solder joint.

After you have the first positive lead in place (Figure **K**), repeat the same steps to connect the negative lead from the same ESC to the negative circuit on the PDB — in our case, the right-hand strip (Figure **L**).

Do the rest of the ESCs in the same manner (Figure **m**). Think about where you will position the wires for all of them. Don't make the mistake of cutting one of your wires too short in an attempt to save space. It's better to leave a little extra length at first.

The only thing left to solder is the main battery lead, which is attached exactly the same way as the ESC leads. Clip some insulation from the end, tin the wire, hold it in place with the pliers, and apply heat. Be sure that you're connecting to the right circuits and that your solder joints are nice and solid.

6. MOUNT BRUSHLESS MOTORS

Brushless motors for small drones such as the Little Dipper are constantly evolving. Due to this shifting landscape, we're not going to give specific instructions for one particular model of motor, but rather the overarching ideas that apply to all the different models.

In general, clockwise-spinning motors should have a reverse-threaded shaft. Counterclockwise motors work best with a standard thread. If you need to confirm the direction that your motor will be spinning, do so before attaching it to the frame.

One catch: Not all manufacturers make their motors this way. Many use only a standard thread — this will still work fine. Whichever style you have, be sure to tighten it well and check it regularly.

Start your mounting process by laying the motor flat on the top of the boom while lining up the mounting holes in the bottom of the boom with the threaded holes in the bottom of your motor (Figure **n**). Make sure you have a motor with the correct thread direction for the anticipated motor direction. The motor leads should run down the length of the boom. Make sure they don't run in any other direction.

PAY ATTENTION TO THE THREAD DIRECTION!
Notice that your motors have a shaft that holds the prop in place. This shaft is threaded, and a nut of some type (called the prop nut) fits over that thread and puts pressure on the prop. In the early days of small drones, those threads were all standard clockwise threads. But because quadcopter motors spin in both clockwise and counterclockwise directions, manufacturers realized that if they created motors with standard and reverse threads, they could use the prop's spin direction to help keep it tight. Make sure you always have a thread that screws on in the *opposite* direction from that in which the prop will be spinning.

Now manually feed the first screw through the boom's motor mounting hole and into the threaded holes on the bottom of the motor. Once the first one is done, feed the opposite screw in, and keep going until you have all 4 in place.

Once you have all the screws manually fed into place, tighten them turn for turn on opposing screws (Figure **O**), similar to tightening lug nuts on a car.

Now that you have the first motor mounted in place, move on to the remaining 3.

7. CONNECT BRUSHLESS MOTORS

Your brushless motors will connect to the speed controllers via the 3 black wires you attached your bullet connectors to earlier. If you're new to brushless motors, you may notice something funny at this point: The wires are not labeled. That's because there is no wrong way to connect a brushless motor to a speed controller, only different directions of motor rotation. You can connect those 3 wires in any possible combination and it would never damage the motor; it will simply spin in one direction or the other.

Our goal is to make the NE and SW motors spin in a counterclockwise direction. But because our build is not complete yet, just hook them all up the same and you can test them later and make any necessary changes — made easier by the bullet connectors between the components.

NO PROPS FOR YOU!
Be sure the propellers are *not* attached to the motors yet. That will be our very last step after we have confirmed everything is working as expected. This is an important safety measure.

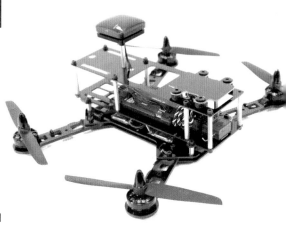

8. CLEAN UP

At this point, it's a great idea to use a couple of zip ties and tidy your wiring job. It's also helpful to place identifying tabs on the servo leads out of the speed controllers before you button everything up. This will make your life much easier when working on subsequent projects later in the book. We usually use a fine-tip marker or paint pen and label the motor number on the ESC lead.

WRAPPING UP

Now that your speed controllers and motors are permanently mounted, make sure everything is firmly attached. Also check all soldering in your power harness/power distribution board to make sure nothing is loose. The connection between your motors and speed controllers is temporary at this point — you'll adjust it later when all the flight electronics are installed. ●

For the complete build, including the next section on adding the autopilot and battery, as well as facts, tips, and history on drone flying, pick up a copy of *Getting Started with Drones* from makershed.com

Special Section
DRONES ◆ BUYER'S GUIDE

makezine.com/drone-revolution

2016 DRONE FLYER'S GUIDE

8 CAPABLE QUADCOPTERS, TESTED AND REVIEWED

Written by Matt Stultz

ON A SUNNY SPRING MORNING, A TEAM FROM *MAKE:* **HEADED** up the hillside of the beautiful Kunde Family Vineyards in Sonoma, California to fly a variety of quadcopters and compare them side-by-side for our 2016 drone guide.

SETUP
With the wide array of machines now available, we realized it was time for our first drone shootout. Focusing on camera-carrying quadcopters that range from $500 to $1,400, we applied the same methodology used with our 3D printer and digital fabrication reviews. We acquired brand-new rigs to compare the unboxing and setup experience, as well as ensure the same new-user look across all the systems. We installed apps and firmware updates on a picnic table in the shade of an oak tree, tethering to our cellphones for the necessary connectivity. (In real use, always check for updates before heading into the field.) And then we took to the air.

TESTING
Our flight tests evaluated battery life, image quality and stability, flight distance, air speed, and handling. For battery life and image assessment, we filmed aerial time-lapse video of a fixed point to gauge movement while measuring overall flight time. For distance, we flew over the property's grape vines and trails to determine where reception would drop. The air speed test involved measuring the elapsed time going from a hover to a point 400 feet away on multiple runs, then averaging the results. Drone handling leveraged the shared experience and expertise of the testers.

RESULTS
We learned a lot through the tests — including that manufacturer claims were almost always higher than real-world performance. And crashes were minimal and non-significant, showing the systems' increased heartiness.

Overall, it was a great day of flying and data collection with just the perfect number of close calls to keep it all interesting. Now, on to the results.

Taking a DJI Inspire 1 for a joy ride after our tests. If you want to go further and faster than the rigs in our guide, this is our dream drone.

MORE ONLINE! Get all the details of how we tested and reviewed our roundup of drones at makezine.com/go/2016-drone-guide

Special Section
DRONES ◆ BUYER'S GUIDE

DJI PHANTOM 4
OBSTACLE DETECTION AND AUTONOMOUS MODES MAKE THIS HIGH-QUALITY QUAD FUN TO FLY
Written by Matt Stultz

DJI HAS BEEN BUILDING A REPUTATION AS THE NAME IN CONSUMER CAMERA DRONE AIRCRAFT. Balancing both affordability and cutting edge, there is a DJI drone for almost everyone. The Phantom 4 is not just the newest arrival in the DJI line of products, it feels like an important evolution for the field.

SENSATIONAL SENSORS
The real standouts on the Phantom 4 are its sensors. Two distance sensors and five cameras, including two above the landing gear, couple with a dedicated processor to detect obstacles in its flight path. This works in all flight modes except the high-speed "Sport" setting. We were skeptical, but it performs well, and is ideal for autonomous flights or when using the new person-following function to help prevent collisions with trees, buildings, or other obstructions along the way. It's just for forward-moving flight, though — the other sides are still blind.

Setup is easy; the new clip-on propeller mounts are brilliant, and the compact styrofoam case is great for transport. The included controller feels comfortable, and advanced functions are easy to navigate on your smartphone or tablet with the DJI app.

CONCLUSION
Advanced pilots and newcomers alike will find the Phantom 4 to be a great flyer and well deserving of its next-gen reputation.

2016 GUIDE TO DRONES — Make: BEST OVERALL — DJI PHANTOM 4

TEST SCORES
	0 1 2 3 4 5
FLIGHT STABILITY	
VIDEO QUALITY	
RANGE	
EASE OF FLIGHT	
CONTROL INTERFACE	
BATTERY LIFE	

Total 28.5

dji.com

- **PRICE AS TESTED** $1,399
- **INCLUDES** Drone, transmitter, 2 sets of propellers, battery and charger, Lightning and microUSB interface cables, carrying case
- **DRONE SIZE** 15.5" diagonal (without propellers)
- **CONTROLLER TYPE** Hybrid: transmitter for control, smartphone or tablet (not included) for live view, telemetry, and mission planning
- **CARRYING CASE** Compact foam case with latch and handle
- **CAMERA TYPE** Built in
- **CAMERA RESOLUTION** 4K

MEET THE TESTERS (left to right)
- **Kelvin Lam**, sales director for drone-racing organizers rotorsports.com
- **Mike Senese**, *Make:* executive editor
- **Matt Stultz**, *Make:* contributing editor
- **Justin Kelly**, founder of 3D design service Proto.house See his project "FPV Night Racing" on page 38
- **Tyler Winegarner**, *Make:* video producer (not pictured)
- **Audrey Donaldson**, *Make:* senior buyer (not pictured)

PRO TIPS
The Phantom 4 provides numerous flying modes. Consider your flight goal before taking off — following a target or flying a specific course might be easier with assisted modes instead of free flight.

WHY TO BUY
The myriad of sensors on the Phantom 4 makes flying not only fun but also less stressful. Now losing communications has a much higher possibility of a safe outcome and recovery.

PHANTOM 3 PROFESSIONAL
GET STARTED IN AERIAL CINEMATOGRAPHY WITH THIS SIMPLE AND STABLE DRONE

Written by Tyler Winegarner

WITH ITS RECENT PRICE DROP, THE PHANTOM 3 PROFESSIONAL MIGHT BE THE BEST STARTING POINT for anyone seriously interested in aerial cinematography. Out of the box, the quadcopter is straightforward to set up. The propellers thread onto color-matched screws, and a wrench in the included toolkit helps achieve critical tightness. After charging the batteries, installing the DJI Go app, and updating the firmware, you're ready to take to the skies.

POWERFUL AND SOLID
The flight experience is simple but very powerful. The controller feels solid with direct and responsive sticks. The GPS and optical positioning system offers rock-solid hover stability, even with poor GPS reception. Its stability yields incredible confidence, even to novice pilots, and the telemetry and Lightbridge-powered live view make it easy to plan and execute tracking shots, even if you're coming up with them as you go along. The 4K video is clear and distortion free, and the AVCHD QuickTime files make the footage easy to add to any production workflow.

CONCLUSION
The Phantom 3 Professional may not be DJI's flagship model anymore, but there's still plenty to like, especially as a starting point in aerial cinematography.

TEST SCORES
FLIGHT STABILITY
VIDEO QUALITY
RANGE
EASE OF FLIGHT
CONTROL INTERFACE
BATTERY LIFE
Total 27.5
dji.com

- **PRICE AS TESTED** $999
- **INCLUDES** Drone, 2 sets of propellers, transmitter, battery, charger, toolkit
- **DRONE SIZE** 15.5" diagonal (without propellers)
- **CONTROLLER TYPE** Hybrid: transmitter for control, smartphone or tablet (not included) for live view, telemetry, and mission planning
- **CARRYING CASE** No
- **CAMERA TYPE** Built-in
- **CAMERA RESOLUTION** 4K

PRO TIPS Being grounded by a necessary firmware update is a huge bummer. Make sure you're ready to fly before an outing.

WHY TO BUY Clear, long distance high-definition video transmission, professional-caliber 4K video recording, and a rock-solid positioning system.

PHANTOM 3 STANDARD
A HIGH-QUALITY QUADCOPTER AT AN AFFORDABLE PRICE

Written by Tyler Winegarner

2016 GUIDE TO DRONES — Make: BEST FOR STARTERS — DJI PHANTOM 3 STANDARD

THERE'S A POPULAR QUOTE ABOUT THE VALUE OF TOOLS: "Good equipment ain't cheap, and cheap equipment ain't good." Neither part of this quote applies to the Phantom 3 Standard.

Just under $500, this model is the entry point in the Phantom lineup. Released a few months after the Professional model, you'd be forgiven for looking at what's missing. Yes, there are a few features absent, but unless you're comparing them side-by-side, you may never notice.

The Standard only uses the U.S. GPS satellites for positioning, and drops the visual positioning entirely. Hover stability suffers a little, especially in wind, but the aircraft remains incredibly easy to control. Live video and telemetry is achieved via a Wi-Fi downlink instead of DJI's Lightbridge technology, and you'll see some latency and frame drops, especially in a saturated network area. The view is clear enough to set up a shot, but you should hesitate to use the live view alone to navigate through close obstacles. Recorded video footage is stable and clear, even if it loses the 4K resolution.

CONCLUSION
Aside from its cheap-feeling controller and weak mobile-device clip, everything about the Standard smacks of quality. If you're on a budget and need a camera in the sky, the Phantom 3 Standard offers incredible value.

TEST SCORES
FLIGHT STABILITY
VIDEO QUALITY
RANGE
EASE OF FLIGHT
CONTROL INTERFACE
BATTERY LIFE
Total 24
dji.com

- **PRICE AS TESTED** $499
- **INCLUDES** Drone, transmitter, 2 sets of propellers, battery, charger
- **DRONE SIZE** 15.5" diagonal (without propellers)
- **CONTROLLER TYPE** Hybrid: transmitter for control, smartphone or tablet (not included) for live view, telemetry, and mission planning
- **CARRYING CASE** No
- **CAMERA TYPE** Built in
- **CAMERA RESOLUTION** 2.7K

PRO TIPS The included gimbal locking bracket isn't stellar. If you have a 3D printer, download and print a better one — Thingiverse has plenty of options.

WHY TO BUY If you need a camera in the sky and you're on a budget, the Phantom 3 offers excellent bang for your buck.

Special Section — DRONES ◆ BUYER'S GUIDE

3DR SOLO

Written by Mike Senese

IF YOU WANT TO GET UNDER THE HOOD AND TWEAK SOFTWARE, THE SOLO HAS YOU COVERED

THE FUTURISTIC, BLACK BODY OF THE SOLO HIGHLIGHTS HOW ADVANCED DRONES HAVE BECOME — the quadcopter and its controller each house a 1GHz Linux computer to handle its autonomous flying features. The autonomy was easy and intuitive to use; our multi-point cable cam route was accurate and repeatable throughout the testing. Based on the open-source ideals from a key player in the DroneCode Project, much of Solo's system is available for the user to modify and share, and will be familiar to the original DIY Drone community members.

FLIGHT PERFORMANCE

The Solo is quite zippy; its slowest setting is great for minimizing jerky motion when filming. But at full throttle, the Solo tied the Phantom 4 (in "Sport" mode) for fastest rig in our roundup. Our battery tests gave 14 minutes of flight time, and we were able to get about 2,600 feet away before losing reception.

3DR's use of a GoPro for the Solo's camera used to be a positive based on GoPro's ubiquity and quality. Competing systems, however, have more or less caught up, and some prefer a flat-horizon look to GoPro's wide-angle lens.

TRAVEL SMART

The Solo's optional backpack was our favorite drone case of all the units in our review. Comfortable and with ample storage for all accessories, it greatly improves the usual hassle of transporting your system to remote locations.

CONCLUSION

Solo is a smart system with future expansion options. But it's not cheap.

3dr.com

TEST SCORES 0 1 2 3 4 5

- FLIGHT STABILITY
- VIDEO QUALITY
- RANGE
- EASE OF FLIGHT
- CONTROL INTERFACE
- BATTERY LIFE

Total 25

◆ **PRICE AS TESTED** $1,149 with optional gimbal and case (+$499 for GoPro Hero 4 Black)
◆ **INCLUDES** Drone, gimbal, transmitter, battery, charger, 2 sets of propellers, travel case
◆ **DRONE SIZE** 18" diagonal (without propellers)
◆ **CONTROLLER TYPE** Hybrid: transmitter connects to smartphone or tablet (not included)
◆ **CARRYING CASE** Tested with optional backpack
◆ **CAMERA TYPE** GoPro cameras (not included)
◆ **CAMERA RESOLUTION** Variable

PRO TIPS
As always, be sure to update the firmware before heading out to the field; with the Solo, this means getting the GoPro firmware updated as well.

WHY TO BUY
You want a quality, steady aerial platform with a software set that you can hack on. And you have a GoPro camera to put on its gimbal.

GHOST 2.0
Written by Matt Stultz

FOR SMARTPHONE-FAVORING PILOTS, THIS MACHINE GOES A LONG WAY

SMARTPHONES ARE AT THE CENTER OF HOW WE INTERACT WITH TECHNOLOGY TODAY. The Ehang Ghost 2.0 takes advantage of our familiarity by making yours its controller. This isn't a new concept, but Ehang's inclusion of the G-Box takes it to a whole new level.

INCREASED COMMUNICATION
Instead of directly connecting a smartphone to the drone, Ehang uses the G-Box as a man-in-the-middle device. The device greatly increases the communication range over the phone's Bluetooth or Wi-Fi signal. We had some initial difficulties getting the G-Box connected during our testing, but once established, it seemed to work well. But, you're still stuck flying without sticks.

When purchasing a Ghost 2.0, you have a few choices for your camera — from none to a built-in 4K unit. The unit we tested had a multi-axis gimbal with a GoPro mount. Our aerial footage showed some video stability issues that we think might have to do with gimbal problems, but this does not seem to be the usual case as we have seen stable Ghost 2.0 video from others.

CONCLUSION
The Ghost 2 is trying to skew the trends in the camera drone world with its upside-down mounted propellers and smartphone control, but it might need a few more tweaks to help its standing in the pack.

TEST SCORES — Total 17 — ehang.com
- FLIGHT STABILITY
- VIDEO QUALITY
- RANGE
- EASE OF FLIGHT
- CONTROL INTERFACE
- BATTERY LIFE

- **PRICE AS TESTED** $799
- **INCLUDES** Drone, G-Box, battery, charger, two sets of propellers, propeller guards, toolkit
- **DRONE SIZE** 16" diagonal (without propellers)
- **CONTROLLER TYPE** Smartphone or tablet (not included) with a range extension box
- **CARRYING CASE** No
- **CAMERA TYPE** GoPro or optional built-in camera
- **CAMERA RESOLUTION** Varies on selection

PRO TIPS The Ghost 2.0 comes with a built-in 5.8GHz video transmitter compatible with most FPV gear. Add a pair of goggles to feel like you're flying.

WHY TO BUY Ehang's unique controller setup makes the Ghost a good step up for those who are already accustomed to flying small drones with their smartphone.

CHROMA
Written by Matt Stultz

THIS STABLE AND EFFORTLESS RIG IS A LOT OF DRONE FOR THE MONEY

THE CHROMA FROM HORIZON HOBBY IS A GREAT DRONE THAT WON'T BREAK THE BANK. While it may not have the name recognition of units from DJI or 3DR (outside the hobby aircraft world that is), this machine brings a lot to the table.

NO FLIGHT DELAYS
Out of the box, the Chroma has everything you need to get flying; no extra smartphone or tablet is needed. The included remote has a built-in touchscreen interface, saving the user from needing to connect an external device to access advanced functions. The built-in 4K camera and gimbal provide smooth, high-quality video and a first-person video feed back to the controller.

Flying the Chroma is a breeze, and while it may not be the zippiest (even with its oversized brushless motors), it's extremely stable and effortless. If there is one downside to the Chroma's performance, it would be the range of the transmitter. While we were able to fly other drones in this class out to nearly a mile away without signal loss, the Chroma lost communication less than 800 feet from the transmitter.

CONCLUSION
If you are looking to start flying on a budget but still want excellent video capabilities, the Chroma is a great pick.

TEST SCORES — Total 22 — horizonhobby.com
- FLIGHT STABILITY
- VIDEO QUALITY
- RANGE
- EASE OF FLIGHT
- CONTROL INTERFACE
- BATTERY LIFE

- **PRICE AS TESTED** $800
- **INCLUDES** Drone, transmitter, extra props, battery, charger, toolkit, screen guard
- **DRONE SIZE** 18" diagonal (without propellers)
- **CONTROLLER TYPE** Standalone with built-in touchscreen
- **CARRYING CASE** No
- **CAMERA TYPE** Built in
- **CAMERA RESOLUTION** 4K

PRO TIPS The charging time on the Chroma's batteries is pretty substantial. Buy extra batteries to help keep your drone in the air and reduce wait time between flights.

WHY TO BUY The Chroma might not be top-of-the-line but it's a great aircraft for those looking to get into the hobby.

Special Section
DRONES ◆ BUYER'S GUIDE ◆ FLIGHT CONTROLLERS

PARROT BEBOP 2
A SLIGHT INCREASE IN SIZE GIVES A HUGE INCREASE IN FLIGHT TIME
Written by Mike Senese

TEST SCORES 0 1 2 3 4 5
- FLIGHT STABILITY
- VIDEO QUALITY
- RANGE
- EASE OF FLIGHT
- CONTROL INTERFACE
- BATTERY LIFE

Total 17.5 — parrot.com

- ◆ **PRICE AS TESTED** $549
- ◆ **INCLUDES** Drone, battery, charger, 2 sets of propellers
- ◆ **DRONE SIZE** 12" diagonal (without propellers)
- ◆ **CONTROLLER TYPE** Tablet- or smartphone-only (tested) or SkyController (optional)
- ◆ **CARRYING CASE** No
- ◆ **CAMERA TYPE** Built-in with software video stabilization
- ◆ **CAMERA RESOLUTION** 1080p

THE BEBOP 2 SIZES UP SLIGHTLY FROM THE ORIGINAL PARROT BEBOP with wider arms, larger propellers, and a bigger fuselage. The main beneficiary of this is the battery, which provided a group-best 27 minutes of flight in our tests — the only rig that surpassed its listed figure.

DEVICES WON'T SUFFICE
We received a controller-less unit for our testing. Parrot states that a separate SkyController will pair with it, but we were unable to get ours to connect and had to fly it using our iPad and iPhone. This meant we were only able to fly as far as the devices would allow, which isn't far at all. We then found ourselves waving the tablet overhead to reconnect. Also, receiving a phone call is a very scary interruption that causes you to lose access to the controller app mid-flight.

The app provides ample feature control for video and photo. Data is stored internally with 8GB of space, not quite enough for a full day of filming. The video quality is that of the original Bebop, suitable for impressive home videos, but not for television or movie production. Autonomous flight planning can be unlocked with a $20 in-app purchase.

CONCLUSION
We like the size of the Bebop 2, but Wi-Fi control continues to be frustrating.

PRO TIPS
Bring your laptop to download video and clear space.
To get the Bebop's full potential, skip the standalone version and get one with the SkyController.

WHY TO BUY
If you need the absolute longest flight time possible, rather than control reliability and image quality, this is for you.

PARROT BEBOP 1
SLEEK, SMALL PACKAGE BUT WI-FI LEAVES MUCH TO BE DESIRED
Written by Mike Senese

TEST SCORES 0 1 2 3 4 5
- FLIGHT STABILITY
- VIDEO QUALITY
- RANGE
- EASE OF FLIGHT
- CONTROL INTERFACE
- BATTERY LIFE

Total 16 — parrot.com

- ◆ **PRICE AS TESTED** $499
- ◆ **INCLUDES** Drone, transmitter, 3 batteries, charger, 2 sets of propellers, tablet sunshade, neck strap
- ◆ **DRONE SIZE** 10.75" diagonal (without propellers); SkyController: 14.6×9×7.5"
- ◆ **CONTROLLER TYPE** Tablet- or smartphone-only or SkyController (tested)
- ◆ **CARRYING CASE** No
- ◆ **CAMERA TYPE** Built-in with software video stabilization
- ◆ **CAMERA RESOLUTION** 1080p

THE SMALLEST QUADCOPTER IN OUR ROUNDUP, THE ORIGINAL BEBOP is compact enough to slip into a normal backpack with the propellers attached, yet not so small that its flight suffers overly. Add the bulky but necessary SkyController with its gargantuan Wi-Fi antenna, however, and the portability advantage drops greatly.

CONNECTION REJECTION
The Wi-Fi control interface is what really hurts this quad. It takes frustratingly long to pair and drops out frequently mid-flight, causing the Bebop to stop and hover until reconnected. Parrot boasts a range of 1.4 miles, but even at top speed, there's no way to fly that far with the dismal battery life, especially if you want to get it back.

Part of the Bebop's size advantage comes from tucking a high-resolution camera and ultra-wide angle lens into the body, and using software to crop the view down to 1080p to provide video stabilization. It works surprisingly well, but is tricky to get slow, smooth pans while in flight. We'd like to have an option to output the entire scene at full resolution.

CONCLUSION
With its size and price, the Bebop is a family friendly quadcopter. The propellers aren't deadly, and its video output is enjoyable. Too bad about the Wi-Fi woes.

PRO TIPS
With its dismal flight time, spare batteries are crucial for any serious use. Buy as many as you can if you can find them.

WHY TO BUY
Get this if you need something small and sleek, and can be patient about getting into the air.

makezine.com/drone-revolution

CHOOSING A FLIGHT CONTROLLER
Written by Alex Zvada

8 DIY BRAINS TO GET YOUR BIRD OUT OF THE NEST

ALEX ZVADA grew up around aviation and is the air-to-air chase pilot and art director at Flite Test.

WHEN BUILDING A DIY MULTICOPTER, ONE OF THE BIGGEST CONSIDERATIONS IS THE BOARD THAT MAKES THE RIG FLY, CALLED A FLIGHT CONTROLLER. This piece translates the movements of your transmitter's sticks into motor movement via the electronic speed controllers (ESCs). It includes onboard gyros that provide stabilization and compensate for outside forces such as wind gusts.

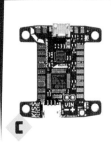

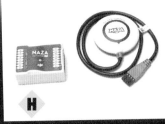

RACING CONTROLLERS

The most popular and affordable flight controllers are the **Acro Naze32** A, **Lumenier LUX** B, and **Flyduino Kiss FC** C. Compact and lightweight, they mount easily to 180-size and larger FPV racing frames.

From a beginner's standpoint, there is not much of a difference between these units. All provide auto leveling, but lack GPS and position hold. Auto level is good for beginners as it automatically snaps the craft back to level when you let go of the control sticks. However, you will still need to compensate for wind/drift and manage your throttle at the same time.

The Naze32 has been around the longest and has the most information available online. Serious/professional racing and freestyle pilots commonly use the LUX and Kiss FC. These boards range in price from $30 to $45.

On a typical multicopter, the receiver connects to the flight controller, which is wired to the ESCs and then to the motors. The fairly new **Graupner GR-18** D combines the receiver and flight controller into a single part, eliminating excess wires and making setup easier.

Like the other three, the GR-18 only provides auto leveling, which is configured through the transmitter's touchscreen LCD, making setup and tuning a breeze. The flight controller and receiver combo make its $99 price very reasonable. Note, however, that you can only use it with a Graupner transmitter.

Considered the grandfather of flight controllers, **HobbyKing's KK2.1** E was the first that boasted an affordable $30 price point (now available for $19) and was somewhat easy to use thanks to a built-in screen. Though the technology is fairly outdated, it is another option for beginners, and still used on many projects today.

AUTONOMOUS CONTROLLERS

Flight controllers with more autonomous features increase in cost to about $200. The most popular are the **Eagle Tree Vector** F, the **3DR Pixhawk** G, and the **DJI Naza-M Lite w/GPS** H. These give you the comforts of GPS and position/altitude hold — if you let go of the control sticks, the craft will hold steady, even in the wind!

The Naza-M Lite is the most user friendly and easy to set up/tune. The Vector has the benefit of a color onscreen display (OSD), which, like a fighter jet, displays vital flight information on your FPV goggles or screen. The 3DR Pixhawk is open source and by far the most programmable of the three; if you are a coder, you can go in and change the parameters of the flight control board to get it to do what you want.

Whatever route you take, there is an affordable and accessible flight controller option for you.

Special Section DRONES ◆ FPV NIGHT FLYING

FPV NIGHT FLYING

HACK AN INFRARED FLOOD LAMP AND RIP THROUGH THE DARKNESS!

Written by Justin Kelly

JUSTIN KELLY is a professional Maker (proto.house) and an avid rapid inventor with a flair for style. His 3D-printed Nerf blaster mods business (lasergnomes.com) was featured in *Make:* Volume 50.

MATERIALS
» **Infrared-capable mini video camera** Sony Super HAD II ($15–$35) or RunCam Owl ($40–$65)
» **Infrared security flood lamp** such as CMVision IR30 or Univivi U03R
» **Battery, lithium polymer (LiPo), 3S or 4S** for the flood lamp. You'll power the camera from your existing FPV power.
» **Battery power lead** to match your LiPo battery

TOOLS
» Soldering iron
» 3D printer (optional)

◆ **TIME REQUIRED:** 1–2 Hours
◆ **COST:** $100–$200

EVERY SECOND COUNTS WHEN YOU'RE FLYING FPV (first-person view), from reaction time and video latency to simply finding the time to fly. Who wants to quit just because the sun goes down?

Here's how to rip through the darkness, flying at night using infrared (IR) cameras and illumination. You can combine popular retail solutions with some Maker innovation to squeeze every last enjoyable moment out of your aircraft.

SEEING IN THE DARK

1. CHOOSE A LOW-LIGHT, INFRARED-SENSITIVE CAMERA

If your camera can't see in low light, you can't fly. As dusk gives way to night, outdoor ambient light fades from 100 lux down to as low as 0.0001 lux on a moonless night. That's quite a range of darkness to conquer.

I flight-tested 2 low-light, IR-sensitive cameras to gauge their nighttime effectiveness; both are small and lightweight and accept a wide range of power, 5V–24V DC (from a 2S, 3S, or 4S battery). Special thanks to Rainer Von Weber (vondrone.com) and SF Drone School for their assistance!

Sony Super HAD II, 600TVL, 28mm×28mm
◆ IR blocking (film on sensor) and IR sensitive versions available
◆ Min. illumination: 0.01 lux (e.g. full moon)
◆ Day/night: Auto/Color/B&W (resolution is increased to 650VTL in B&W mode)
◆ White balance: Manual/Auto/Auto track
◆ Features: 2.1/2.8/3.6mm lens options, digital noise reduction, wide dynamic range, OSD setup menu and controller

RunCam Owl, 700VTL, 19mm×19mm
◆ IR blocking: on lens, removable
◆ Min. illumination: 0.0001 lux (starlight)
◆ Day/night: Auto
◆ White balance: Auto
◆ Features: ½" f/2.0 150° wide-angle lens, voltage regulator, onboard mic

RunCam

Sony's 28mm mini cam is an FPV standard, and the IR-sensitive version worked well at night with the proper illumination. But RunCam's Owl (Figure **A**) was designed for FPV night flying, and was the hands-down winner in my testing.

TIP: If your IR-blocked Sony has its IR film on the CCD sensor (not the lens), you can remove it to make the camera IR-sensitive.

2. INSTALL YOUR IR CAMERA
a. Remove your existing FPV equipment.
b. Replace with your IR-sensitive camera: Yellow wire is signal, red is power, and black is ground. I created a generic 3D-printed housing for mounting the Sony camera on my drone (Figure **B**); you can download the file for printing at tinkercad.com/things/75mSEWpClDy. The Owl is so small you can just use its existing housing.
c. Before powering on, check wiring for polarity and make sure your power line is providing the right voltage for your camera.
d. Confirm that the camera is functioning, then proceed to camera setup if available. The RunCam Owl, having no setup or controller, is basically done once installed.

The Sony has an in-depth on-screen display (OSD) setup menu allowing you to really tune your camera to your local conditions. For night flying, there are a few settings to adjust to gain the most from the camera while reducing latency. The most important is to force it to B&W while reducing the shutter speed to 1/60.

INVISIBLE FLOODLIGHTS
So you've flown around and you like what you see. But you don't like what you still can't see — the areas too dark for even the keenest of vision systems.

Let's burn off the veil of darkness with infrared illumination. Infrared light is so far to the red end of the light spectrum that humans can't even see it. But we can employ technology to give us bionic, superhuman vision by simply removing the camera's IR filter and employing an IR light source.

WHY IR AND NOT JUST A FLASHLIGHT?
Infrared light travels farther and takes less power to generate, so it's great for mobile platforms. And IR is undetectable to the naked eye, so it won't disturb anyone with beams of light flying around in the night.

3. CHOOSE A DONOR INFRARED ILLUMINATOR
I sourced 2 different IR flood lamps designed for night vision security systems, then gutted the electronics for mounting on my aircraft. Both systems run on 12VDC, emit infrared light at 850nm, and accept traditional 3S and 4S batteries without complaint. (Direct wiring is possible but not recommended, as the invisible lights will drain your flight battery to unsafe voltage levels if you forget about them. Instead, I add a small 3S battery.)

CMVision IR30 Wide Angle IR Illuminator
◆ 30 high-power IR LEDs, 50' range

Univivi U03R 90° IR Array Illuminator
◆ 4 large high-power IR LEDs, 100' range

I tested both and found that the properly aimed 4-pack generated more focused light at a distance outdoors. The 30 smaller LEDs would likely be better for indoor flying where speeds are lower and lateral illumination is important. Your usage may vary. You can also experiment with a higher-voltage battery to gain more IR light.

4. MAKE YOUR INFRARED DRONE FLOOD LAMP
a. Remove the IR array from its metal housing (Figure **C**).
b. Cut the provided 2-pin power lead, then join it to a new power lead for an isolated battery (recommended) or for existing aircraft power (Figures **D** and **E**).
c. Install the IR array onto your aircraft so that its angle matches the angle of your FPV camera. You can download and 3D print my 2"-diameter Flood Lamp Mount from tinkercad.com/things/1uyb4KSgN6h.
d. Power up your IR array. Here you can see the Sony camera's view with ambient light only (Figure **F**) and then with the 30-LED infrared lamp switched on (Figure **G**) — huge difference!

You're ready to pierce the darkness. Remember, when flying at night, even with the aid of technology, it's best to know your surroundings and be familiar with obstacles that are hard to see even in the best conditions. Now go rip!

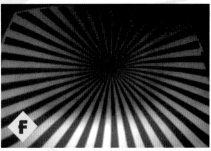

See more photos and share your night flying FPV tips at makezine.com/go/infrared-night-flying.

Special Section
DRONES ◆ PILOT PROFILES

Asa Hammond, Niels Joubert, Nathan Schuett, and Zoe Stumbaugh

THESE MULTIROTOR VANGUARDS ARE PUSHING THE FIELD TO NEW HEIGHTS
TOP GUNS

Written by Mike Senese

WITH UAVS QUICKLY DEVELOPING INTO HIGHLY CAPABLE FLYING PLATFORMS, countless applications for them have appeared spanning entertainment to industry. Each category now has notable pilots pushing their fields even further. We caught up with a few of the standouts to hear how they got started, what they're flying, and where they're going next.

MIKE SENESE is the executive editor of *Make:* magazine. He spends his spare time tinkering with remote-control aircraft and attempting to cook the perfect pizza.

NIELS JOUBERT
Drone researcher, Stanford University

Many of the functions drones are starting to offer today — such as basic autonomy and obstacle avoidance — had been developed and demonstrated for years in research labs. That practice continues now, with academics around the world designing quadcopter feats like throwing and catching objects, weaving rope bridges, and stacking blocks. Stanford Ph.D. student Niels Joubert is part of this, focusing on reproducing computer-graphic movements in the real world. See his work at njoubert.com.

◆ **WHEN DID YOU GET INTO DRONES?**
I started Stanford Computer Graphics' Quadrotor project in September 2013, along with my colleague Mike Roberts. That was the start of my research foray into drones. Before that, I worked on other mechatronic projects, including the SNAPS Cubesat.

◆ **WHAT DOES YOUR INVOLVEMENT IN THE DRONE SPACE MEAN?**
In our research, we almost exclusively fly fully autonomous missions. We're building new tools and algorithms that use quadrotors purely as a tool for placing sensors in the world without a human in the loop. We're also exploring ways the drone controls itself, interacting with people in much more fluid and natural ways than a person holding a controller.

In my personal life, I love doing aerial cinematography and casual FPV flying. With the Stanford UAV Club, we have a great time chasing each other, flying very different vehicles!

◆ **WHAT'S IN YOUR FLIGHT CASE?**
- A 3DR Solo with an integrated Swift Navigation RTK GPS
- A Black Pearl FPV monitor
- An RTK-GPS base station: NovAtel GPS antenna, Swift Navigation RTK GPS, and a Ubiquiti Bullet M5 Wi-Fi radio
- A MacBook Pro ground station running MAVProxy and my custom open source experimental flight environment, "Spooky" (github.com/njoubert/spooky)
- A couple of GoPros and tripods to record the flight from the ground
- A small fold-up table and chair
- A big tarp — sometimes you need to protect against unexpected rain!

◆ **WHERE WILL DRONES GO NEXT?**
I see drones heading in two major directions: swarms of drones working together fully autonomously to perform tasks such as monitoring or package delivery, and drones as personal robots, making it easier for us to capture and share our lives.

ZOE STUMBAUGH
FPV Racer

In the new sport of FPV racing, the best have only been flying for a few years, as the technology is still so young. Zoe Stumbaugh is one of the stars who has emerged, developing new flying techniques and featuring prominently in last summer's National Drone Racing Championships in Sacramento, California — the first sanctioned national race of its type.

◆ **WHEN DID YOU GET INTO DRONES?**
I bought my first "Nano" drone in August of 2014, and started work on my first FPV craft in September of 2014 after watching videos on YouTube. Fell in love with flight on the first battery!

◆ **WHY DO YOU FLY?**
I fly for the sheer fun of it — and the thrill of competition. Besides racing I've been pioneering "3D" flying, where the machine has the ability to reverse thrust near instantly, allowing for maneuvers that visually defy gravity. See makezine.com/go/zoe-fpv.

◆ **WHAT'S IN YOUR FLIGHT CASE?**
My backpack is overloaded, it carries four rigs: a Twitch 109 from UniqueFPV for everyday fun and racing, and setups from Hovership Zuul, Atomic Aviation Mercury, and Bullit Drones as my 3D test machines. My Fat Shark HD v2 goggles, Taranis 9XD transmitter, 30 or so batteries. Big-bag-o-props. A 5.8GHz DVR/LCD screen for passengers and a spare set of goggles for giving rides. GoPro and Mobius HD cameras for capturing footage.

◆ **WHERE WILL DRONES GO NEXT?**
Drones will become fully autonomous with sense and avoid technology for most applications, with human input "guiding" things. In the racing scene I'm expecting full HD video out for live broadcast on national television by the end of the year.

NATHAN SCHUETT AND ASA HAMMOND
Co-founders of Prenav.com

Industrial inspection is a particular challenge for large structures — especially those that move, like wind turbines. To do a proper inspection, an engineer traditionally would need to ascend the tower with climbing gear in order to get close access to crucial areas. The use of UAVs with high-resolution cameras could eliminate the danger that these inspectors face, but current units do not have the repeatable precision needed to fly to the exact spot reliably. Prenav, co-founded by Nathan Schuett and Asa Hammond, aims to overcome that by tracking their inspection drones inside a point cloud to guide their position within a centimeter of accuracy.

◆ **WHY DO YOU FLY?**
Mainly we're drawn to the tech. We find these tiny aerial robots to be fascinating — especially the challenge of enabling them to fly with autonomy and precision.

◆ **WHAT'S IN YOUR FLIGHT CASE?**
Prenav prototypes (our laser scanning guidance system and drone).

◆ **WHERE WILL DRONES GO NEXT?**
Longer flight times, better imaging capabilities, increased safety and reliability, and more intelligent collision avoidance and automation. Can't wait!

NANCY EGAN
Regulatory expert and 3D Robotics general counsel

Over the past year, the FAA has released a much-anticipated set of rules around consumer drone use, including a pilot registration requirement that turned out much easier and agreeable than many anticipated. To make this happen, the agency organized a task force comprising leading drone experts, manufacturers, and vendors, who developed the current directives. Nancy Egan, general counsel for 3D Robotics, was a lead voice on this project, and continues to be engaged with regulators in Washington as the field develops.

◆ **WHAT DOES YOUR INVOLVEMENT IN THE DRONE SPACE MEAN?**
One of my primary jobs is to help people at all levels of government understand the technology and the speed of innovation, and help shape regulation and policy that fosters innovation — while keeping us all safe.

◆ **WHY DO YOU FLY?**
I fly in Sonoma. I am a casual videographer. Sonoma is absolutely gorgeous. Because of the weather and terrain, you can fly the exact flight plan every day at the exact same time and you get a new kind of gorgeous every time.

◆ **WHAT'S IN YOUR FLIGHT CASE?**
A print out of my FAA registration!

◆ **WHERE WILL DRONES GO NEXT?**
Is it cliché to say the sky is the limit? In the short term we need to make sure we have a legal framework in place in the U.S. to allow people to use the amazing technology we already have.

Special Section
DRONES • DO-IT-YOURSELF DRONES

IT WASN'T LONG AGO THAT THE ONLY WAY TO GET YOUR HANDS ON A QUADCOPTER WAS TO BUILD ONE YOURSELF. Today, even with all the great drones you can fly right out of the box, there's still a lot to be said for making your own and understanding it inside and out. Here are four buildable rigs that caught our eye.

BUILD YOUR

GET AERIAL WITHOUT BREAKING THE BANK BY MAKING

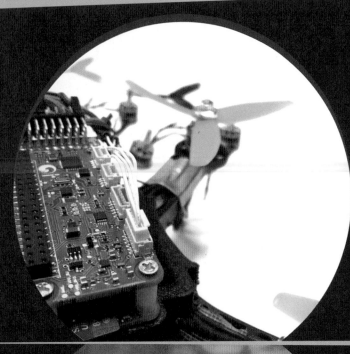

PI ZERO DRONE

This quad packs a Raspberry Pi Zero for extra onboard smarts. The Linux-based drone takes advantage of Dronecode's open source APM software allowing for both lightweight tasks, like stabilization, and advanced autopilot tasks (following, circling, scripted missions, hold, return-to-home, and other modes). Estimated bill of materials is around $200, $60 of which is spent on the HobbyKing Spec FPV250 Kit that serves as the mechanical foundation of the build.

DETAILS:
- FPV250 frame kit
- PXFmini autopilot board
- Multistar 1704 motors
- 1000mAh Turnigy 11.1V (3S LiPo) battery
- 5×3×3 propellers
- Afro V3 12A ESCs
- Mini JST power connections

BUILD IT:
makezine.com/go/pi-zero-drone

MICRO 105 FPV

The two-piece design of this micro quadcopter is small enough to be printed on just about any 3D printer. The drone uses a $33 Micro Scisky flight controller board and has a space carved out in front for a compatible $10, 170° FPV camera. Total weight (without battery) comes in at around 38 grams. The bill of materials is under $100.

DETAILS:
- Micro Scisky flight controller
- Hubsan X4 H107c 8.5mm motors
- FX758-2 5.8G 200mW video transmitter
- 170° wide angle camera
- Hubsan X4 battery
- Ladybird props
- JST SH connector
- M3 nylon nuts and 20mm machine screws (cut to size after attaching to the frame)

BUILD IT:
thingiverse.com/thing:1221911

makezine.com/drone-revolution

OWN DRONE

Written by Donald Bell

DONALD BELL is a former projects editor for *Make:*. He enjoys playing guitar, skateboarding, and hanging out with his son.

THESE QUADCOPTERS AND FIXED-WING R/C PROJECTS

FIELD-STRIPPABLE 3D PRINTED QUAD

Racing drones break often. In fact, if yours isn't breaking down regularly, you're probably not pushing it hard enough. To minimize downtime, Andrew Ocejo from Ozone Drones designed a modular quad airframe that assembles in less than 5 minutes with no tools required. A handful of hitch pin clips are the only non-3D printed parts you'll need to get up and running.

DETAILS:
- Emax 2300kV motors
- 5×6 carbon fiber props
- Emax 12A ESCs
- 1400mAh 3S battery
- CC3D, NAZE32, or APM mini flight controller

BUILD IT:
makezine.com/go/field-strippable-quad

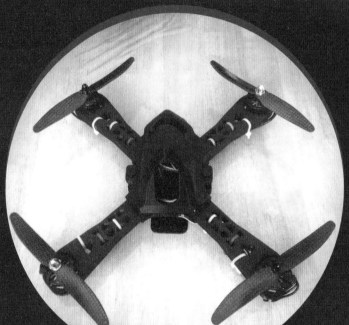

FLITE TEST MIGHTY MINI SPARROW

The Flite Test crew has built an extensive and impressive roster of DIY quads and fixed wings. This little guy is a fast and cheap way to get started building your own single-motor R/C plane. The foamboard-based design invites rapid prototyping, and can be used as a motorless glider, or fully built-up into an agile FPV racer with a nose-mounted camera.

DETAILS:
- Weight without battery: 6.5oz (187g)
- Wingspan: 28.5" (723mm)
- 2300kV 1806 size motor
- 5×3–6×3 prop
- 12A ESC minimum
- 800 mAh 3S battery minimum
- 5-gram servos

BUILD IT:
flitetest.com/articles/ft-sparrow-build

Special Section   DRONES ◆ WI-FI SECURITY RISKS

Written by Brent Chapman

ANTI-DRONE WI-FI HIJACKER

BUILD A PI-POWERED DRONE DISABLER TO UNDERSTAND THE SECURITY RISKS OF WIRELESS COMMUNICATIONS

BRENT CHAPMAN is an active duty cyber warfare officer in the U.S. Army assigned to the Defense Innovation Unit Experimental (DIUx) in Silicon Valley. When not in uniform, you can find him in his wood shop or basement tinkering and building.

QUADCOPTERS CAPABLE OF TRANSMITTING HIGH-QUALITY VIDEO are making it possible to affordably record unique perspectives. But these "unmanned aircraft systems," as the FAA calls them, have posed new challenges in security, safety, and privacy, and many experts caution pilots to consider the implications of increased drone usage. In addition to the concern of constant surveillance, there's the possibility that businesses (or hackers) can collect location information from mobile devices by using roving drones.

As a result, a cottage industry is forming for anti-drone technology. These devices come in a range of sizes, from plane-mounted to handheld tools. I will show you how to build our own rig to execute a particular network-based attack against one type of quadcopter control: Wi-Fi.

WHY 802.11?
Wi-Fi is a key interface for many current quadcopters. Some use it as the interface between the controller and a tablet displaying mapping and telemetry data. A

Raul Arias

few drones, such as Parrot's Bebop and AR.Drone 2.0, are entirely controlled via Wi-Fi. This type of system lowers the barriers to entry into the drone space since pilots can use their own devices for control, but it does create interesting security situations since existing network-based attacks can now be used against these devices. Modern drones are essentially flying computers, so many of the attacks that were developed for use against traditional computer systems are also effective. The AR.Drone 2.0 in particular has many impressive features and sensors that users can access, and its low cost makes it an ideal platform for experimentation and learning.

HOW IT WORKS

The AR.Drone 2.0 creates an access point that the user can connect to via a smartphone. The access point that it creates is named **ardrone2_** followed by a random number. This access point by default is open and offers no authentication or encryption. Once a user connects the device to the access point, he or she can launch the app to begin control of the drone. This process, though convenient for the user, makes it easy to take control of the drone. The AR.Drone 2.0 is so hackable, in fact, that there are communities and competitions focused on modifying this particular drone.

OUR TEST

Using a laptop computer, USB Wi-Fi card, and our new antenna (see page 47, "Build a Cantenna"), we'll explore a very simple attack. Power on the AR.Drone 2.0 and have a friend fly it around using the app. After a few seconds, its access point should also show up in your available wireless networks. Connect to the network and start up your favorite terminal application. The default gateway address for this network will have an address of **192.168.1.1**. You'll be able to telnet to this address since the service is, unfortunately, left wide open on this system. Telnet is an older protocol for accessing remote computers. At this point, you can explore the system, or shut it off entirely without the legitimate user knowing what's going on. Using a combination of freely available network tools, you can easily perform all these steps from your computer.

Now we'll look at how you might automate this attack with a Raspberry Pi, a touchscreen, and a couple of Bash scripts.

I used a great tutorial provided by Adafruit (learn.adafruit.com/adafruit-pitft-28-inch-resistive-touchscreen-display-raspberry-pi) to set up my Raspberry Pi with a touchscreen, so that I could launch my attacks with a click. Assuming that you have a Pi already set up, let's walk through how you could automate this.

The first step is to log into your Pi using SSH (Figure **A**).

Change directory to the Pi's desktop (or wherever you want) so that the scripts are easy to find and click (Figure **B**).

Using your favorite text editor, create a new file. I named this *join_network.sh* because I'll be using this to make

A WORD OF CAUTION
While I won't touch on signal jamming or directed energy, it's worth noting that jamming creates serious safety risks and is illegal. Additionally, the computer-based techniques that we'll cover should only be done on networks and devices that you own, or have permission to experiment on.

```
[Ajax:~ brent$ ssh pi@192.168.2.205
[pi@192.168.2.205's password:

The programs included with the Debian GNU/Linux system are free software;
the exact distribution terms for each program are described in the
individual files in /usr/share/doc/*/copyright.

Debian GNU/Linux comes with ABSOLUTELY NO WARRANTY, to the extent
permitted by applicable law.
Last login: Wed Mar 16 02:37:40 2016 from 192.168.2.27
pi@pitft:~ $
```
A

```
[Ajax:~ brent$ ssh pi@192.168.2.205
[pi@192.168.2.205's password:

The programs included with the Debian GNU/Linux system are free software;
the exact distribution terms for each program are described in the
individual files in /usr/share/doc/*/copyright.

Debian GNU/Linux comes with ABSOLUTELY NO WARRANTY, to the extent
permitted by applicable law.
Last login: Wed Mar 16 02:37:40 2016 from 192.168.2.27
pi@pitft:~ $ cd Desktop/
pi@pitft:~/Desktop $
```

Special Section
DRONES ◆ WI-FI SECURITY RISKS

the Pi automatically join the AR.Drone 2.0 access point (Figure **C**).

Add these 8 lines to your script (Figure **D**). On line 7, enter the full name of the AR.Drone 2.0 access point. Once you're done, save everything.

You're now going to automate the connection that you tested before and send an additional command to shut the drone down. Start by creating another script. I called mine *poweroff.sh* (Figure **E**).

Add these lines to your script (Figure **F**). This initiates a telnet connection to the drone, which is located at **192.168.1.1**, and sends the command of **poweroff**, which tells the drone (which is a computer after all) to shut everything down.

Now make sure that the scripts are *executable*. Do this by typing **sudo chmod u+x** *filename*. Check this for both of the files; we can verify that they are now executable by typing **ls -la** and looking for the read, write, execute permissions **rwx** associated with the file (Figure **G**).

The two scripts are ready to use. Be sure that no people or fragile items are below the drone when you're testing. Have fun!

OTHER DRONE-RELATED POSSIBILITIES
This is just the tip of the iceberg — there are a number of things that an attacker could do. These include modifying or deleting system files, intercepting video and sensor feeds, rerouting the drone to alternate locations, or a combination of these. Hacker and Maker Samy Kamkar, the person behind security projects like RollJam and MagSpoof, even released a project designed to allow an attacker drone to autonomously seek out any Parrot drones within Wi-Fi range, disconnect the real user and initiate a new connection that is controlled by the attacker drone. The end result essentially is an army of "zombie" drones.

We also tested a range of drones at the *Make:* office that rely on some form of Wi-Fi connectivity for their operation. All of the drones tested were susceptible to deauthentication (deauth) and disassociation attacks, which forced all users off the drone's access point, resulting in a loss of connectivity to the drone.

BEYOND DRONES
The DIY "cantenna" is incredibly useful for vastly extending range of connectivity. Using the Raspberry Pi rig we've just assembled, an attacker could reprogram the computer to perform a number of attacks, such as a deauth attack against a coffee shop hot spot. How is this useful? Well, consider the following scenario: An attacker sets up a fake access point called "Better Wi-Fi" that is designed to collect credentials. Customers are content using the real coffee shop's connection so there's no reason for them to join the attacker's fake network. Knowing this, the attacker uses his rig to deliver the deauth against the real access point to force all the users off. The users can no longer reach the real access point, and in need of internet connectivity they connect to the evil (but convincing sounding) hot spot and their account credentials are collected.

HOW TO PROTECT YOURSELF
The first step, of course, is educating yourself on the capabilities of your drone, its limitations, and good security practices. There are advantages to using Wi-Fi, for example, as the means to control machines, but there are many things to consider from a security point of view, such as wireless security protocols, encryption, and open ports. For more sensitive applications, there are far more secure options when it comes to command-and-control. Always ask permission and tinker safely! ◉

Written by Brent Chapman

BUILD A CANTENNA
BOOST YOUR WIRELESS SIGNAL WITH A DIRECTIONAL ANTENNA MADE FROM A CAN

makezine.com/drone-revolution

IN A WIRELESS WORLD, CONNECTIVITY IS KING. A good antenna attached to your wireless device will boost your signal and dramatically extend your range. In less than an hour, you can build your own directional "cantenna" to connect to distant wireless hot spots or interact with wireless devices like some of the drones featured in this issue.

1. CALCULATE
The toughest part about this build is calculating the best location for mounting the radio connector, and the correct length of the wire element for ideal performance of the antenna. Fortunately, there are lots of online resources to help you with the math, such as csgnetwork.com/antennawncalc.html. Figure A gives an overview of how the measurements are calculated.

Given the dimensions of the can, about 100mm in diameter, the Type N connector needs to be mounted 44mm from the bottom of the can. The frequency we're interested in is in the 2.4GHz band, so the total height of the copper wire needs be roughly 31mm.

2. MEASURE AND MARK
Measure 44mm up from the bottom of your can, and mark the position for the N connector with a permanent marker (Figure B). (I measured 44mm down from the top of my cookie tin, which has a replaceable lid that I used as the back of my antenna.)

3. DRILL HOLES
On the mark you made, drill a hole so that your N connector can fit snugly. It's good to start with a small bit and work your way up until the hole is just large enough. Once you're done, sand the area around the hole to ensure good contact with the connector.

Test-fit the connector and mark the 4 mounting holes (Figure C). Drill these to match the machine screws you'll use to mount the connector. Or skip the screws and just solder the connector to the can.

4. SOLDER THE WIRE TO THE N CONNECTOR
You need to prepare the connector before it's mounted. Take a 4" piece of straight copper wire — the straighter the better — and remove any coatings.

Now you're going to solder that short copper wire to the top of the connector. It's a little tricky; I used helping hands to position everything before soldering it in place (Figure D).

After soldering the wire to the connector, test-fit again and then trim the wire to the distance you calculated in Step 1. In my case, that was 31mm.

5. MOUNT THE CONNECTOR
If you didn't solder the connector to the can, tighten the machine screws from the outside of the can into their nuts inside. If needed, you can cut the bottom off the can for access, then tape it back in place when you're done. There — you have a brand new cantenna (Figure E).

6. CONNECT TO WI-FI CARD AND ENJOY
Screw the pigtail cable into your card and the N connector. Your cantenna is ready to use (Figure F).

GOING FURTHER
You can add a coat of paint to make it more tactical, or add a handle or mount it on a tripod for precise aiming.

MATERIALS
- » **Metal can, 3¼"–4" diameter, open at one end** 3½" diameter is ideal. The Pepperidge Farm Pirouette cookie tin works well at 3.875".
- » **RF connector, chassis mount, Type N female** Amazon #B009PL6BD0
- » **Pigtail cable, Type N male to RP-SMA male** to connect wireless card, Amazon #B003U6825G
- » **Bare copper wire, 12–15 gauge, 4" length**
- » **Machine screws and matching nuts (4)** to mount your Type N connector to the can

TOOLS
- » Soldering iron and wire cutter
- » Drill
- » Can opener
- » Sandpaper, 320–400 grit
- » Helping hands tool (optional)
- » **External wireless card** such as Alfa AWUS036NHA USB wireless adapter

◆ **TIME REQUIRED:** 1 Hour
◆ **COST:** $10–$20

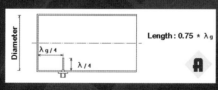

Skill Builder
TIPS AND TRICKS TO HELP EXPERTS AND AMATEURS ALIKE

SERVOS 101
Learn to make things move with this simple motor
WRITTEN BY EMILY COKER

EMILY COKER is the former *Make:* Labs coordinator. She is a Jill of all trades who loves drawing and reading comics, and can cook like a mad scientist.

Servomotors are used in everything from toys and drones to household items like DVD players. They come in a variety of configurations (1 , 2 ,and 3 are the most common); understanding the differences is the trick to figuring out which is right for your project.

In simple terms, servos are standalone electric motors that push and rotate parts in machines wherein a specific task and position need to be defined. When a servo is given a command, it moves to a position and holds there with resistive force. A servo uses either a rotary actuator or a linear actuator to control angular or linear positions through acceleration and velocity. They typically operate between 4.5V–6V, which run through power, ground, and control wires.

A servo's DC motor rotates at a high RPM on low torque. The gears inside the servo convert the output to a much slower rotation speed but with more torque. This creates a large amount of force for a short period of time and is a perfect example of the basic law of physics:
Work = force × distance.

1
POSITIONAL ROTATION SERVO
This variety rotates within a 180° range. It's not designed to turn beyond its preset limits. Useful for limited-range applications like moving levers or steering linkages.

SERVOS AND PWM
A servo contains a small circuit board with a sensor that communicates its rotational position to an R/C controller, computer, or microcontroller. The resulting information is then translated into small electrical pulses with variable energy. Manipulating this energy, called pulse width modulation (or PWM), controls the position of the motor.

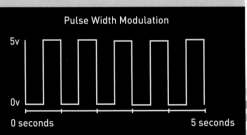

WHAT'S INSIDE

The internal components on a servo motor are what give it such useful versatility. Standardized case sizes and interchangeable accessory brackets **A** minimize design complexities while providing interface options for almost any application. The small DC motor **B** keeps the servo size minimal, while the controller board **C** monitors motor position and offers user interface and control. A speed-reducing gear set **D** allows for precision motor alignment with high levels or torque.

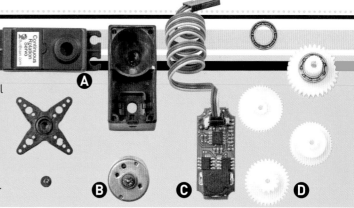

2. CONTINUOUS ROTATION SERVO

While levers are often used on standard servos, wheels and gears become more useful with this style, which can turn in any direction independently and continuously.

3. LINEAR SERVO

This type offers more gears than the positional rotation servo, but is otherwise very similar. It uses a rack and pinion mechanism to change the output back and forth instead of circularly. This servo is rare, but can be found in larger hobby planes and robots.

ANALOG VS. DIGITAL SERVOS

Analog and digital servos look exactly the same. The difference is in the way they signal and process information.

Analog Servos operate based on on/off voltage signals that come through the PWM. When this type of servo is resting, the PWM is essentially off unless you transmit an action. Producing torque from the resting mode makes the initial reaction time sluggish, which can cause problems in advanced R/C applications.

Digital Servos use a small microprocessor to receive and direct action at high frequency voltage pulses. The digital servo sends 300 pulses per second, where the analog only operates at 50 pulses per second. These faster pulses provide consistent torque for quicker and smoother response times. This is a great benefit, but digital servos consume a lot more power.

Skill Builder
Wet Sanding

POLISH WITH WET SANDING

Remove large scratches from shaping and make your project shine

WRITTEN BY JORDAN BUNKER

When you really want a glossy finish, you'll need to wet sand. Wet sanding is a process that's often used on car paint jobs, guitars, and even 3D prints to give them a mirror-smooth look.

Wet sanding is typically done after dry sanding to get an even finish. Unlike dry sanding, wet sanding is done not to shape a surface, but to remove the large scratches left by dry sanding. When done correctly, the surface will be slowly leveled out, and the scratches left behind will become smaller and smaller, until the light they reflect no longer makes them visible.

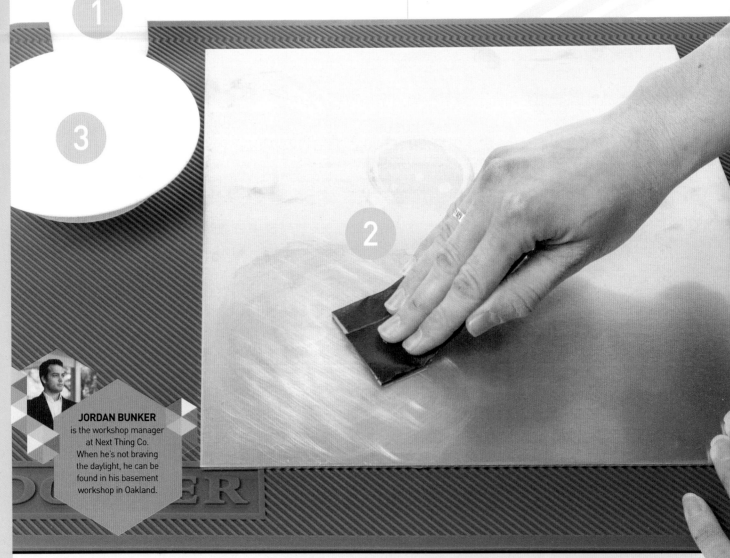

JORDAN BUNKER is the workshop manager at Next Thing Co. When he's not braving the daylight, he can be found in his basement workshop in Oakland.

1

The wet part of wet sanding refers to the use of water or some other liquid as lubrication to help carry away grit particles that are removed. Without the liquid, material can build up in the sandpaper and leave behind scratches that are larger than the particle size, ruining your finish. In general, the best liquid to use for most materials is water with a little bit of detergent in it. (Dish soap works well.) The detergent lowers the surface tension of the water, and helps wet the paper and the material more thoroughly, reducing scuffing. If you're sanding bare metal, you can use WD-40 as lubrication instead of water.

2

Not all sandpaper is created for wet sanding, so make sure that the sandpaper you're using is specifically rated for it. 3M's "Wetordry" is the standard type, and can be found at your local hardware store. You can fold the paper over on itself to create a thick piece to hold on to, but it's even better to wrap the sandpaper around a backing pad. Alternatively, you can buy sanding sponges which have the abrasive adhered directly to the sponge. This helps the sanding surface conform to the shape of the material.

3

In order to thoroughly wet the sandpaper, many people recommend soaking it in liquid overnight. This ensures that the paper won't absorb any more moisture during the sanding process. If you don't have time to soak the paper overnight, aim for at least 15 minutes of immersion prior to sanding.

QUICK TIP

A wise Maker once told me the key to sanding is to forget about the concept of "sanded." Don't get impatient! It can be boring, but it's important to be thorough. Put on your favorite show, or listen to your favorite music, and really get into the process.

WHAT IS "GRIT?"

The grit of a sandpaper refers to the size of the abrasive particles bonded to the paper. The higher the grit, the smaller the particles, and the finer the scratches left behind.

The first grit that you start wet sanding with depends on the previous grit you used to shape the object, and how smooth the surface is. If you last dry sanded with 600 grit, you'll want to choose a wet sanding grit that's around 800–1200. In general, you'll want to start wet sanding around 600–1200 grit, and follow the same dry sanding rules for working up through the grits, jumping up 200–500 grit between passes (depending on how meticulous you want to be). You can buy wet sandpaper up to about 3000 grit, but most people will be satisfied around 1500 or 2000 grit.

One big difference between dry sanding and wet sanding is the movement used. Dry sanding requires small circles; wet sanding uses straight lines, alternating direction between passes. This way, each successive pass works to remove the scratches from the previous one. Be sure to use a light touch as well — we're not trying to remove a lot of material, just the scratches!

PROJECTS | Music Visualizer Table

Get Your Freq on

Written by Charlie Turner

Hack a $10 Ikea table with LEDs to make a jumbo music visualizer that dances to your tunes!

CHARLIE TURNER is a 35-year-old father of two, who works in IT but has a passion for anything he can make in his garage, be it robots, furniture, or electronics. He is currently building a life-sized R2-D2, much to the dislike of his wife

Time Required:
A Weekend
Cost:
$80–$110

Materials
- **Ikea Lack side table, 21⅝"×21⅝"** Ikea model #200.114.13, about $10 from ikea.com
- **Arduino Uno microcontroller board** A smaller Arduino-compatible board would probably also work.
- **Adafruit Bicolor LED Square Pixel Matrix with I2C Backpack** Adafruit Industries #902, adafruit.com
- **Electret microphone with amplifier** Adafruit #1063
- **LEDs: red (64) and green (64)** Green plus red will produce a third color, amber.
- **Acrylic sheet, translucent or frosted, 16"×16"** aka plexiglass or perspex. About $15 from TAP Plastics or eStreet Plastics. You might also find it at home improvement stores.
- **Hardboard or strong cardboard** for creating the grid
- **White paper or paint** to cover the hardboard
- **USB cable**
- **Hookup wire: black, red, and green**
- **Jumper wire, male-male**
- **Breadboard**
- **Parchment paper (optional)**

Tools
- Drill
- Handsaw
- Rotary tool with cutoff wheel (optional) such as a Dremel
- Large ruler
- Wire strippers The "self-adjusting" kind will save you a lot of time!
- Soldering iron
- Utility knife
- Hot glue gun
- Computer with Arduino IDE software free download from arduino.cc/downloads
- Multimeter (optional) helpful for troubleshooting connections

NOTE: The little bicolor LED display on the Adafruit I2C board doesn't get used in this project, but you can't buy the board on its own. As this does all the I2C decoding, I feel it's well worth the $16. You could also just use a couple of MAX7219 chips — but that would require more wiring, and rewriting the supplied code.

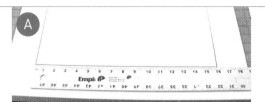
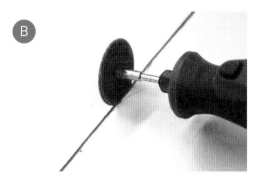

TIP: First drill a hole inside one corner of your square, or make a preliminary cut on your line using a Dremel. Then carefully cut the whole square out with a small handsaw, utility knife, or Dremel.

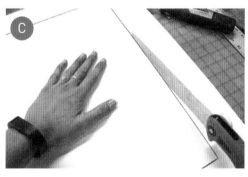

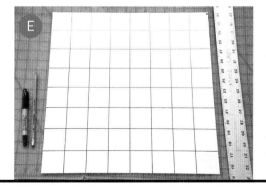

HERE'S A TABLE I BOUGHT FROM IKEA AND INSERTED LOADS OF LEDS, ELECTRONICS, AND A MICROPHONE INTO, so I can put it next to the stereo and have a display that reacts and "dances" to the music played. It's great for parties, and it makes a superb conversation piece. It's got an Arduino for a brain and uses a minimal amount of current, so you can run it off an iPad charger or any USB port that's handy.

Plug it in and watch each column of LEDs jump to the bass, midrange, and treble frequencies like a jumbo-sized graphic equalizer or spectrum analyzer. I based it on the Tiny Music Visualizer project from Adafruit (learn.adafruit.com/piccolo), using their I2C multiplexer board for a tiny 8×8 bicolor LED matrix. The Arduino code is from there, the circuit is from there — all I really created was a big handmade LED matrix, and put it into an Ikea table!

I did troubleshoot a couple of glitches for you — I found that if I connected the circuit as described by Adafruit, with a common ground for the microphone and the display, the results were erratic. If you connect the circuit as I'll show here, using separate grounds on the Arduino board, it works fine. I also found that Chinese Arduino Nano clones did not run the code properly. Switching to the full-sized Uno solved this problem.

BUILD YOUR OWN MUSIC VISUALIZER TABLE

1. MODIFY THE TABLE
Start with an Ikea Lack table, about $10. The matte white was on sale, so that's what I used. Draw a square in the center of the tabletop (Figure A) — mine was 16"×16" (40cm×40cm), giving me room for an 8×8 matrix of 2" (5cm) square sections.

Carefully cut out the square from the tabletop and set it aside (Figures B and C).

Remove the corresponding 16"×16" square of honeycomb cardboard inside the table and discard (Figure D). This will create a space about 1¹¹⁄₁₆" deep for installing the electronics.

2. MOUNT THE LEDS
You'll use your tabletop cutout as the mounting board for your LEDs. First, mark the big square off into 64 little 2" squares, in an 8×8 grid (Figure E).

In each square, mark and drill holes for 2 LEDs: one green, and one red. Put them pretty close together so their colors will blend nicely to amber when they're both lit. To speed up marking, you

PROJECTS | Music Visualizer Table

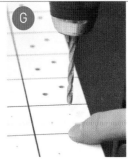

can make a little template from a scrap of cardboard or hardboard. I placed green toward the upper left and red toward the lower right (Figures F and G).

Fill the holes with the LEDs — 64 red and 64 green — inserting them from the back so they emerge on the white top (Figures H and I). Take a picture with all the LEDs in place. Feel good about what you've done.

3. WIRE UP YOUR LED MATRIX

You're just making a very large version of the common cathode matrix shown in the wiring diagram (Figure J). The numbers here indicate which pins you'll connect to on the Adafruit LED controller board.

First, cut some wires and strip them, lots of them — I suggest you buy a good wire stripper, I did and it made it so much easier! Once you think you've got enough, cut some more: You'll need 56 black, 56 red, and 56 green wires. Each should be about 3" (7cm–8cm) long (Figure K), but this will vary depending on your array, so just check before you cut all of them.

Next, to save my fat fingers whilst soldering, I twisted the wires into chains of 7 — end to end — so I could solder them easily without trying to hold multiple wires at once (Figure L).

On the back of the LED board, bend the legs of each pair of LEDs so that their cathodes touch, but their anodes are separate. Solder all the cathodes together using the black wires (Figure M), following the wiring diagram.

Solder all the anodes of the red LEDs together using the red wires. Then solder all the anodes of the green LEDs together using the green wires (Figure N). Finally, cut and solder 24 wires, one to the end of each row and column. These will run off-board to connect to the electronics (Adafruit board, breadboard, and Arduino) so their lengths will vary. It's easiest to locate the electronics near the bottom edge of your LED panel (the front of the table) but you could choose another location, just make sure these off-board wires are long enough to reach.

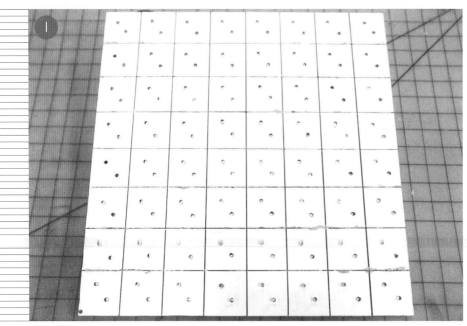

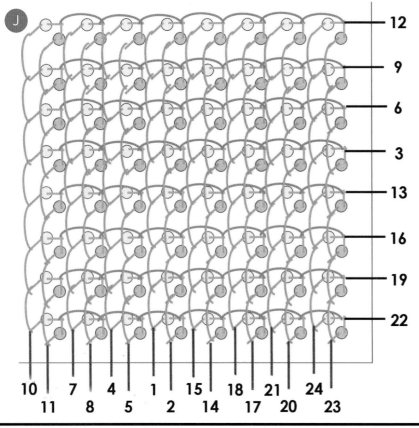

TIP: You could use red for all the anodes, but we found using red and green here makes the connections a lot easier to follow later. —*Jenny Ching*, Make: Labs

54 makershed.com

4. CONNECT THE ELECTRONICS

Connect your newly created LED array to the Adafruit Bicolor LED "backpack" board by soldering its 24 wires directly to the top of the circuit board, following the wire numbering from the LED wiring diagram (Figure J) you used in the previous step. As shown here in Figure , the backpack board's pin numbers run clockwise from 1 at top right to 24 at top left.

Now plug the microphone and the LED backpack board into your breadboard and use jumper wires to connect the Arduino as shown in Figures P and Q.

NOTE: I've mainly used the circuit diagram from Adafruit, but I've made one change: The GND wire from the microphone goes to a different GND connection than the LED array. I found that if I used the same common GND for both, the microphone output was corrupted and the display behaved erratically and didn't respond to music at all. If you use 2 different GND pins on the Arduino Uno, everything works fine.

5. PROGRAM THE ARDUINO

Download the Adafruit_LED_Backpack library from github.com/adafruit/Adafruit_LED_Backpack and put it in your *Arduino/libraries* folder. Download the project code from github.com/adafruit/piccolo, and move the *ffft* folder to your libraries folder too.

Open the sketch *Piccolo.pde* in the Arduino IDE on your computer, then upload it to your Arduino board. Test the matrix and circuit by playing some music (Figure R). You should see the columns of LEDs jumping and dancing to the sound!

6. BUILD THE MATRIX GRID

Cut 18 strips of hardboard or cardboard approximately 15½" long and 1⅛" wide. This width is based on the depth of the table, so test-fit your LED panel and electronics into the table before deciding how wide to cut your strips.

On 14 of the strips, use a saw to cut intersection slots every 2", to half the depth of your strip (Figure S, following page).

For high reflectivity, I used some white laminated hardboard I found in the garage. Then I used the kids' "gloopy glue" (PVA white glue!) to stick reflective photo paper to the back of the hardboard, and cut it to

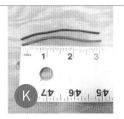
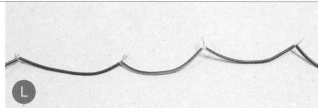
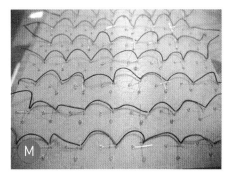
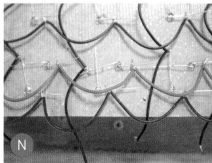

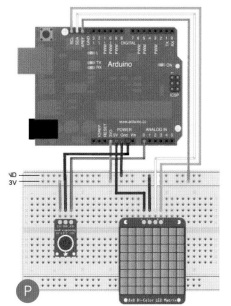

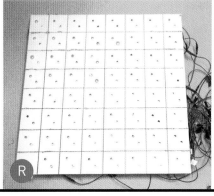

PROJECTS | Music Visualizer Table

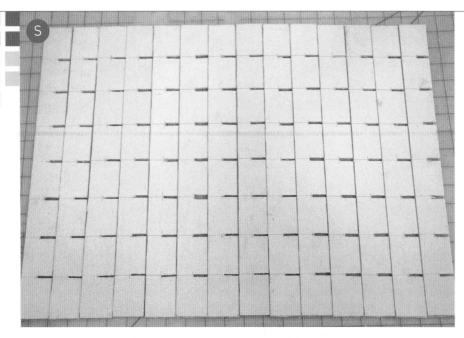

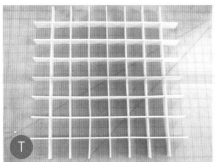

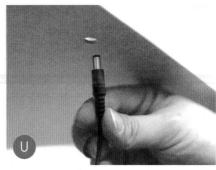

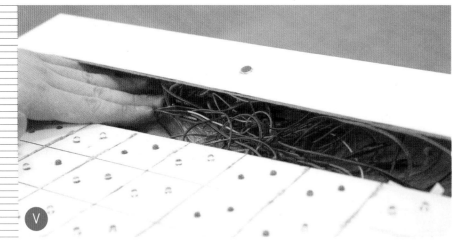

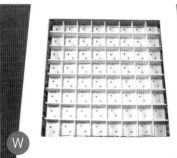

size with a craft knife once dry. (It's probably easier to just paint everything white.)

Now assemble the grid and hot-glue the strips in place (Figure T).

7. COMPLETE THE TABLE

Get a piece of plexiglass (perspex) the same size as your matrix. I covered mine in frosted window film, but if you can find translucent or frosted plexi in the first place, it will save you some effort. For added diffusion, I used some baking parchment paper as well.

Cut a little hole in the bottom of the table to allow a USB cable to fit through (Figure U). Hollow out a pocket in the honeycomb cardboard to house your electronics.

Now slide the electronics into place (Figure V). You can leave the microphone inside, or mount it poking out of the table someplace you like. When it's all aligned to your satisfaction, you can tack the grid down to the LED panel with a few blobs of hot glue (Figure W).

Finally, glue the plastic top in place. I used a hot glue gun, and squeezed a thin bead between the acrylic and the tabletop (Figure X). It's not substantial, but it keeps everything situated.

Your table is now complete!

GET THIS PARTY STARTED

You and your Music Visualizer Table are ready to party. Plug its USB cable into a phone charger or any USB port, then crank up the tunes and enjoy watching the giant tricolor pixels dance like splashing fountains of spectrum analyzer goodness (Figure Y)!

HOW IT WORKS

» **Audio Analysis and Graphics** The microphone you're using has a built-in op-amp that provides a nice clean amplified signal. The Arduino code reads the mic input on Analog pin 0 about 9,600 times per second (9.6kHz) and performs some fancy math called a *fast Fourier transform (FFT)* to convert the raw audio samples into a frequency spectrum. Additional fancy math is performed to split this spectrum into 8 separate frequency bands and analyze the intensity level of each band.

Each of the 8 levels is then rendered graphically from 1 to 8 pixels high — low levels are green, medium are yellow, and high are red. These pixels are redrawn zillions of times a second and sent to the

LED board. It even does that cool trick where the peak pixel hangs for an instant before falling back down (we love that!).

If your mic's not picking up enough sound to really rock the LEDs, try mounting it in a more exposed location. (Or just turn the music up!)

» **LED Multiplexing** Adafruit's Bi-Color 8×8 LED matrix "backpack" board is designed to drive their colorful little 2-inch 8×8 pixel matrix (Figure Z) that has 64 red and 64 green LEDs inside, for a total of 128 LEDs controlled as an 8×16 matrix (2 LEDs per pixel). In the Music Visualizer Table, you're simply substituting a bigger, homemade LED matrix that's laid out exactly the same way.

Like most LED displays, this one is "multiplexed" to save wiring — rather than connecting each LED to a microcontroller individually (129 wires!), they're connected to each other in a grid of columns and rows, controlled by just 24 wires. When the microcontroller passes voltage through the correct column and row, an individual LED lights up. Efficient!

» **I2C Communications** But 24 is still a lot of pins — more than your Arduino board has to spare. The Adafruit multiplexer board solves that problem by using a dedicated LED controller chip to drive all the LEDs. All the Arduino has to do is write data to that chip using the 2-wire I2C serial communication interface, a common protocol designed to let one chip talk to another. The Arduino Uno can easily communicate with I2C devices using its dedicated pins for I2C clock signal (SCL) and I2C data (SDA), shown in Figure AA. For older Arduino boards, just use analog pins 5 and 4. Now you're controlling 128 LEDs using just 2 pins on your Arduino.

Adafruit wrote a basic Arduino library for handling the LED multiplexer board. It can be adapted to any I2C-capable microcontroller — just check yours to see which pins it uses for I2C communications.

GOING FURTHER

You can easily program your own **custom graphics and animations** to run on your LED table anytime — just install Adafruit's GFX Arduino library (learn.adafruit.com/adafruit-gfx-graphics-library).

Before I started this project, I really wanted to build a **Tetris table**. How would you go about creating something like that? ⬤

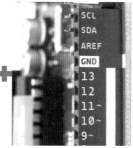

Watch the Music Visualizer Table in action and share your ideas at makezine.com/go/music-visualizer-table.

MORE GREAT BLINKENLIGHTS
at makezine.com/projects

SPARKLE SKIRT
Make clothing that lights up when you move! This is a sewing (no soldering!) project using 12 color-changing NeoPixels and the Flora accelerometer/compass module. From our book, *Getting Started with Adafruit Flora*.
makezine.com/projects/sparkle-skirt-using-adafruit-flora

LUMINOUS LOWTOPS
Take light-up sneakers to the next level with full-color LEDs and force sensors that respond instantly to the pressure of your stomps, jumps, and dance moves.
makezine.com/projects/luminous-lowtops

WORLD CONTROL PANEL
This fun box of switches and LEDs lets your young evil genius light up the world's major cities, custom battle zones, and "home base," complete with sound effects, a Larson Scanner, and "global red alert" mode!
makezine.com/projects/world-control-panel

PROJECTS

makezine.com/51

Custom Catwalk
This light-duty floating shelf hangs mysteriously with no visible support

Written, photographed, and illustrated by Charles Platt

Time Required:
2–3 Hours
Cost:
$15–$25

Materials
» **Pine board, 1×6 or 1×8** your choice of length
» **Steel rods, ¼" diameter, 6" long** Buy 1 for every 16" of board length, McMaster-Carr #8927K18, mcmaster.com
» **Paint or polyurethane**
» **Carpet scrap (optional)** for kitties to dig their claws into

Tools
» **Electric drill and extra-long ¼" bit** such as Sears #00966060000P
» **Speed square**
» **Hammer**
» **Stud finder**
» **Pliers**
» **Carpet needle**
» **Brush or roller** for polyurethane or paint
» **Router (optional)** to round the front edge of the shelf

CHARLES PLATT is the author of *Make: Electronics*, an introductory guide for all ages, and its sequel *Make: More Electronics*. makershed.com/platt

MY CATS LOVE TO CLIMB, SO I DECIDED TO BUILD SOME ELEVATED SHELVES FOR THEM IN OUR LIVING ROOM. Since a cat weighs much less than a stack of books, there was no need for a heavy-duty installation, so I devised a simple, elegant "floating shelf." It's also appropriate for ornaments and framed photos, if for some reason you don't have cats.

Figure shows the concept. Steel rods are partially embedded in the wooden studs in the wall. Holes are drilled in the back edge of the shelf, and the rods fit into the holes. No supporting brackets are needed.

BUILDING THE FLOATING SHELF
At the McMaster-Carr website, I found precut 6" sections of ¼"-diameter steel rod for $1 each. To encourage the rods to slide smoothly into holes in the shelf, I beveled their ends on a grinding wheel, although this may not be strictly necessary.

I drew a horizontal line on the wall, then went along it with a stud finder. To verify the results, I gripped a carpet needle in pliers and pushed it through the sheetrock at small intervals until I found the precise edges of each stud. Then I used an extra-long drill bit to make holes about 3" deep in the stud centers. The long bit made it easier to drill at exactly 90°, because I could place a speed square along the bit while drilling.

I hammered the steel rods tightly into the holes, then laid my 1×8 pine shelf on top of the rods, and marked the location of each rod on the underside of the wood. Now the tricky part: drilling precisely. The speed square was essential.

After rounding the front edge with a router, finishing the board with polyurethane, and stapling a strip of carpet to the top, I came to the anxious moment. Would the shelf fit onto the rods? The trick was to turn it at a slight angle to the wall, to engage the rods one at a time (Figure B). A hammer finished the job, and friction with the rods held the shelf securely.

SUSPENDED IN SPACE
My cats enjoy their elevated perch (you can see one of them in Figure C, waiting for an additional section to be added), and visitors admire the floating shelf while wondering what could possibly support it. One cautionary note: This technique won't support a heavy load, and isn't appropriate for boards wider than 8". ⊘

See more adorable cat photos and share your shelf building ideas at makezine.com/go/floating-shelf.

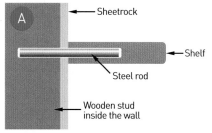

A — Sheetrock, Shelf, Steel rod, Wooden stud inside the wall

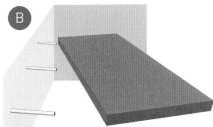

B

C

PROJECTS

makezine.com/51

1+2+3 Rhythm Bones

Written by Phil Bowie ■ Illustrated by Andrew J. Nilsen

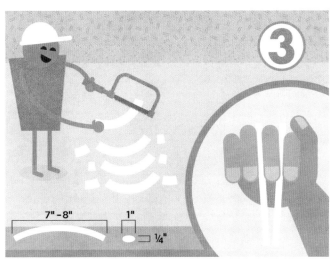

BONES MAY HAVE BEEN THE EARLIEST MAN-MADE MUSICAL INSTRUMENTS, and the skills have rattled down the ages through Egypt, Rome, England, and Ireland to 19th-century American vaudeville and minstrel shows. Rhythm bones still chatter away like country castanets in several genres, including ragtime, bluegrass, and zydeco. Make your own set (traditionally a pair for each hand) and enjoy a hearty meal as well.

1. COOK
At your supermarket, buy beef "short ribs" (not pork or lamb). These are cut from the "plate" section where the ribs are slimmer, well away from the backbone. You want 4 ribs that are 7"–8" long, about ¼" thick at the center cross-section, and 1" wide. Your butcher may think you're being entirely too picky, but ignore the glares and rude remarks. Cook per grandma's instructions. Don't forget to eat your veggies on the side.

2. CLEAN
You could loan the leftover ribs to your dog for cleaning (probably 2–3 hours for a Chihuahua, maybe half an hour for a German shepherd, or 20 seconds for a hungry pit bull), then bleach them in the sun for a week or two. Or you can boil them until scraps clean off easily, then dry them well in a 150°F oven for an hour.

3. CUT
When cool, hacksaw to length, then file or sand rough edges. Carve your initials near one end of each bone.

NOW ROLL THEM BONES. Search for YouTube videos to learn how to play: Dom Flemons (beginner class), James Yoshizawa (expert tutorial), and David Holt (songs and stories). Then take a bow to the applause!

Want to make wooden bones? Get tips at makezine.com/go/123-rhythm-bones

PHIL BOWIE is a lifelong freelance magazine writer. The new, fourth novel in his suspense series is out on Amazon. Visit him at philbowie.com.

Time Required: A Weekend
Cost: $5–$10

You will need:
» Beef short ribs, about 8"×1"×¼" (4) or wood of similar size
» Hacksaw
» File or sandpaper
» Knife

PROJECTS | Building a Mini Tank

Written by Brian Bunnell

Tot-Sized Tank

Your little one will tackle any terrain in this almost unstoppable tracked vehicle

BRIAN BUNNELL is a mechanical engineer by education, but a tinkerer at heart. He earned his engineering degree from Clemson University, and has been working in mechanical design ever since.

Time Required: Lots (but worth it)
Cost: $500–$1,000

MY DAD AND I USED TO BUILD THINGS IN OUR FLORIDA HOME'S 2-CAR GARAGE. Our neighbors always kept watch for the next "Bunnell project," and we were rarely without a crowd when we unveiled a new construction. When I was 4 years old, my dad built me a train that I could get in and drive; it was the coolest thing on the block! It had a trash can for a boiler, PVC sewer pipe for the smoke stack and cylinders, plywood wheels, and it was powered by an automotive windshield wiper motor and a motorcycle battery. It ran on PVC pipe rails, and had a "flat car" to pull my friends along behind me.

I have great memories of that train, and I wanted to build something for my son of a similar magnitude. Initially, I was going to build another train, but then I decided that I wanted to do something different, but equally memorable.

I had always thought having a tank would be fun, and I had toyed with ideas of how one motor could drive 2 tracks independently. I started with the track itself, not really intending to do much with it — it was an experiment to see how I could build a simple track and drive assembly, using readily available and relatively inexpensive materials and components. Once the tracks were successfully complete, the drive assembly, frame, and body organically developed into a more solid idea, and the vehicle was born!

This project is quite involved, but can easily be broken down into easy-to-manage modules.

Materials
- **Aluminum T-slot extrusion**
- **Conveyor guide rail, high-performance polyethylene (UHMWPE)** with stainless steel backing channel
- **Bottle conveyor chain, 104 links** Intralox 880T-K325
- **Lumber: 2×4 and 2×2**
- **Ball bearings, shielded, single-row (8)**
- **Polycarbonate sheet, ½" thick** aka Lexan
- **Aluminum round, 3" dia., 6061-T6 grade**
- **Steel rod, ½" diameter** stainless or mild
- **Pulleys, ½" bore: 4.45" OD (3), 2.05" OD (2), and 1.75" OD (1)** McMaster-Carr part #6204K282, 6204K121, and 6245K120
- **Pulley, ⅞" bore, 2.5" OD** McMaster 6204K137
- **Drive belts, ½" V-belt (4)**
- **Hex bolts, various** with washers, lock washers, and hex nuts
- **Square nuts**
- **Jam nuts**
- **Setscrews, ¼-20**
- **Plywood, ⅜"**
- **Deck screws and wood screws**
- **Metal conduit, ½" EMT**
- **Setscrew conduit couplings**
- **Aluminum angle, 2"×2", ¼" thick**
- **Aluminum bar stock, ½"×1"**
- **Eye bolts (4)**
- **Hanger bolts (2)**
- **Delrin plastic rod, 1" dia.**
- **Aramid rope, 3/16" dia.**
- **Cord pulleys, swivel type, ¾" OD (2)**
- **Perforated steel strap**
- **Cable wire rope clips**
- **Hardware cloth, ¼"×¼"**
- **Mobility scooter batteries, 12V sealed lead-acid (2)**
- **Electric motor, 24V**
- **Switch box**
- **Switches: 3-position (1) and toggle (1)**
- **DC converter, 24V to 12V**
- **Solenoid, 12V**
- **Charging jack**

Tools
- **Reciprocating saw with metal cutting blade**
- **Drill with twist, countersink, and Forstner bits**
- **Vise**
- **Propane torch**
- **Router with straight bit**
- **Miter saw**
- **Electric sander**
- **Rubber mallet**
- **Band saw**
- **Metal lathe** (optional)
- **Threading taps: ¼-20 and ⅜-16**
- **Grinder or mill**
- **Allen wrenches**
- **Pipe bender**
- **Tube cutter**
- **Staple gun**

PROJECTS | Building a Mini Tank

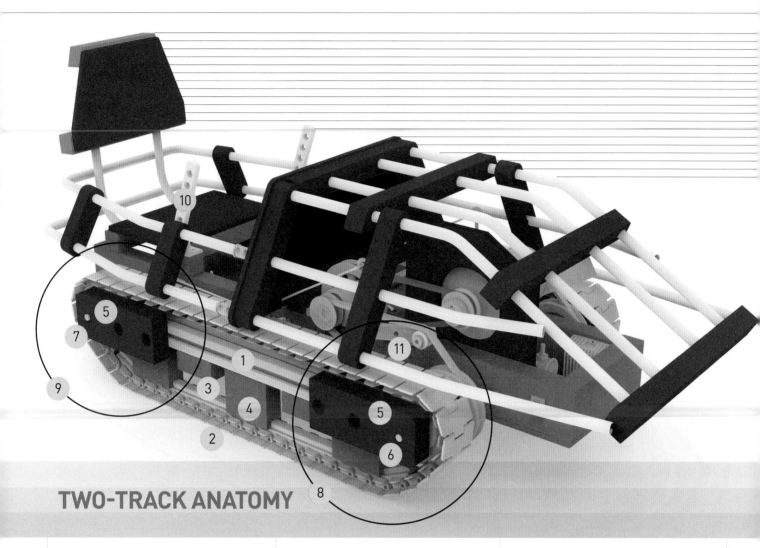

TWO-TRACK ANATOMY

A "tank tread" or "caterpillar" drive has two independent *continuous tracks* that must run at identical speeds for forward travel, but at differing speeds for turning. Here's how I built mine:

1 Backbone
The backbone provides the core structure of each track. I chose aluminum T-slot extrusion because it's rigid and lightweight, and provides T-slots for adjusting tension.

2 Continuous Track
Each track chain is a continuous loop of 52 industrial bottle conveyor chain links.

3 Guide Rail
This serves as a bearing surface for the track chain. It's cut from a conveyor guide rail made of tough, slippery high-modulus polyethylene (UHMWPE) plastic with a stainless steel backing channel. On each end, the steel is trimmed and the plastic bent upward to give the proper approach angle of the chain to the sprockets.

4 Riser Blocks
The wooden riser blocks (3 per track) provide space between the backbone and track chain. Straight grooves are routed lengthwise down the edge of a standard 2×4, which is then cut into 6 blocks (3 per track).

5 Bearing Blocks
The bearing blocks (4 per track, 2 facing in and 2 facing out) secure the drive and idler shafts to the backbone, and are made of standard 2×4 wood and shielded, single-row ball bearings (Figure B). The ends are chamfered for track chain clearance.

6 Drive Shaft and 7 Idler Shaft
The drive and idler shafts are made of ½"-diameter stainless steel rod. Flat sections are machined on the sides to align with the pulley and hub setscrews.

8 Drive End: Drive Pulley and Sprocket
On the drive shaft, there's a sprocket and hub assembly sandwiched between the 2 bearing blocks, and a drive pulley that receives power from the drive belt.

9 Idler End: Brake Pulley and Sprocket
The idler end is similar to the drive end but it's engineered for braking, not driving. Instead of a drive pulley it's got a much smaller idler brake pulley that engages with a fixed strip of V-belt (Figure E, page 64) to slow the vehicle.

10 Track Control Handles
Push forward to engage the drive belt tensioner; pull back to engage the brake.

11 Track Drive Belt Tensioner
The tensioner arm is the key element that engages or disengages the drive belt from the constantly spinning motor (see page 65, "Understanding the Drivetrain").

Brian Bunnell

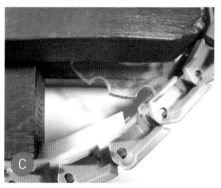
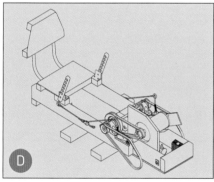

MODULE 1: TRACKS
Building the mini tank is a many-weekend project. Here I'll give a detailed overview; you can follow the complete step-by-step instructions at makezine.com/go/mini-tank.

Assembling the Track Frame
After cutting all parts to size, position the 3 riser blocks so that the grooves are down, and place the backbone on top. Align the front and rear risers with the ends of the backbone and the center riser with the center of the backbone. Use flat-head wood screws to attach each riser block to the backbone.

Now gently (but firmly) press the guide rail into the grooves. Use flat-head wood screws to attach each riser block to the guide rail, ensuring that the screw head is flush with the plastic surface.

Sprocket and Hub Assembly
The sprocket aligns, secures, and transmits torque to the track chain, and is made of ½" Lexan (polycarbonate) — acrylic is too brittle for this application. Sprockets can be purchased, but I designed and cut these to specifically fit the track chain I used. After determining the sprocket size, teeth number, and spacing needed, I made a template and transferred it to the Lexan. I then cut the Lexan using a drill (for the teeth) and a band saw. You need 4 sprockets total (2 per track). The drive and idler sprocket and hub assemblies are identical.

The hub transmits torque from the shaft to the sprocket, and is made of 6061-T6 aluminum with a ¼-20 setscrew. Hubs can be purchased, but I machined these to fit the sprockets and shaft I was using. After determining the bolt pattern needed for mounting the sprockets, I used a metal lathe to machine a flange in the aluminum, and drilled the 4 mounting holes through the flange. Then I drilled the ½" center shaft bore, and drilled and tapped the setscrew hole to the center shaft bore. You need 4 hubs total (2 per track).

Slide a sprocket and hub onto a length of ½" rod, so that the flange is against the sprocket, then mark the 4 hub mounting holes onto the sprocket. Remove the sprocket and drill the marked holes. Align the hub mounting holes to the holes in the sprocket, and from the hub side, push a 5⁄16"×1" bolt through each hole; affix (in this order) a plain washer, lock washer, and standard nut on each bolt, then tighten (Figure A).

Mounting the Bearing Blocks
Use a rubber mallet to tap each bearing tightly into its pocket, so that it's flush with the bearing block surface and doesn't slip.

Lock the hubs to their shafts by tightening their setscrews over the flat sections. Sandwich each hub/sprocket between its bearing blocks (with the bearings positioned inward), then lock the pulleys to their shafts too. Bolt the finished assembly to the backbone using the mounting holes on the bearing blocks (Figure B).

Mount and Tension the Track Chain
Slide the track chain loop onto the track frame, seating the chain in the drive sprocket. Slide the idler bearing blocks toward the center of the track frame, allowing room to slide the chain onto the idler sprocket. Align the teeth of the idler sprocket with the inside of the chain (Figure C).

This vehicle is driven from the fixed (non-tensioning) front end, and tensioned from the rear, non-driven end, which is also where the brake pulley is mounted.

To tension the track chain, use a rubber mallet to tap and slide each idler bearing block toward the rear to create the proper chain tension. Tension is correct when the chain can be lifted about ¼" above the center of the backbone.

MODULE 2: FRAME AND DRIVETRAIN
Frame
The wooden frame provides mounting points for the brake belts, control handles, track drive tensioner, body, tracks, seat, motor, live-drive assembly, and electronics, as well as a compartment for the batteries (Figure D).

Two 2×4s, standing on edge, run the length of the vehicle, set 12½" apart (outside to outside). These are tied together by two 2×4s that are laid flat and ultimately set the track width. My track width is 23".

The front section of the frame is inset within the longitudinal 2×4s, to make motor placement easier. This inset section also provides a front bumper, a spacer for motor alignment, and helps correctly position the vehicle's center of gravity.

The frame is floored on the front and back with ⅜" plywood. The back flooring protects the driver from the ground, and the front flooring protects the motor and electronics, and provides a skid-plate underneath.

All frame pieces are assembled with deck/wood screws.

Seat
The seat's primary structural element is ½" metal electrical conduit (EMT), which is bent

PROJECTS | Building a Mini Tank

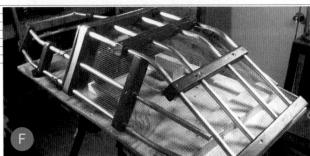

with a hand pipe bender. The conduit is tied together with 2×2s cut to the overall desired seat width, which in this case is the frame width of 12½".

The seat back and bottom are made of ⅜" plywood, and are secured with wood screws to the cross members.

Track Control Handles
The control handles are made from ½"×1" aluminum bar with a ⅜" pivot hole. On the end below the pivot hole, they're tapped to receive an eye bolt. This eye bolt provides a mechanism to adjust the arm's actuation length.

The 4 pivot brackets (for control handles and belt tensioning arms) are made of ¼" thick 2"×2" aluminum angle, cut to 2" long. The "pivot" itself is a ⅜" bolt threaded into a tapped hole in the bracket. You set the drag on the pivot and secure it torsionally using a jam nut on the pivot bolt against the bracket.

Track Drive Belt Tensioner
Each track's tensioner arm (seen in Figure J) is also made from ½"×1" aluminum bar with a ⅜" pivot hole. It is tapped on the top, rearward of the pivot hole, to receive an eye bolt for adjusting the actuation length.

On the front of the arm, parallel to the pivot hole, is a tapped hole to receive a bolt that serves as the axle for the tensioner's small follower pulley. The follower is cut from 1"-diameter Delrin plastic, then machined with a slight dished groove to prevent the drive belt from shifting when the follower is in contact with it.

Brake Belt
The brake consists of a short scrap of ½" drive belt that's affixed to the outside of the frame on the idler end using wood screws (Figure E). This belt must engage with the idler brake pulley, so you'll align and attach it only *after* the track assemblies are mounted to the frame.

Combined Mechanical Control System
The drive handle is linked to the brake and tensioning arm via aramid rope, which is relatively inextensible, yet flexible. The link between the control arm and the tensioner utilizes a single pulley in order to increase the amount of throw of the tensioner arm relative to the throw of the control arm.

To connect the brake belt to the control rope, I used a small scrap of perforated steel pipe strap material. I used a grinder to make flat spots on the belt where I needed attachment, then wrapped the pipe strap material around the cord and belt, and bolted the pipe strap together, pinching the cord and belt in place.

Motor and Live-Shaft Drive Assembly
Whenever the motor is powered, it constantly drives a ½" steel "live shaft," via pulleys and a belt. The live shaft is supported by 2 pillow block bearings mounted to the top of the longitudinal frame members (Figure J). This shaft needs to protrude equally through each bearing so that the left and right track drive pulleys can be mounted and positioned to line up with the drive pulley of each track.

MODULE 3: BODY
Front and Rear Body Structures
The primary purposes of the body are safety and aesthetics. It's built the same way as the seat, with 2×2s supporting a structure made of metal conduit.

The central part of the body is a hollow, open space that we'll call the bulkhead, and is made of ⅜" plywood. A transverse 2×2 is attached to the front, bottom side of the bulkhead and has 2 drilled holes that will act as the body's central mounting point. These holes are centered on the longitudinal frame 2×4s, spaced 11" apart. The bulkhead forms the central support for the body and also aligns the conduit structure into the proper body shape and position.

The front section of the body is completely screened with ¼"×¼" hardware cloth in order to prevent little fingers from accessing the drive systems. The hardware cloth is laid in from the underside and stapled directly to the 2×2s (Figure F).

Dashboard
The dash is made of ¼" plywood, cut to fit within the geometry of the bulkhead. The primary purpose of the dash is safety — to prevent access to the belts and drive system from the driver's area (Figure G).

Any gauges in the dash are completely optional, but they look cool and can be quite functional. I chose to install 3 digital voltmeters (24V, 12V, and 5V), 2 lighted switches (for future use), and a digital temperature sensor to monitor motor temperature.

Below: The electronics responsible for managing power to the motor are housed in the bulkhead of the vehicle. Clockwise from the lower left, you can see the red and black leads to the charging jack; 12V motor solenoid switch; switch module; and 24V to 12V converter.

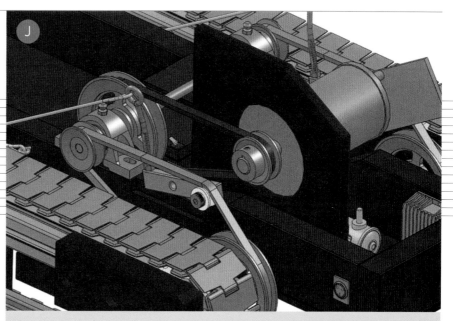

MODULE 4: ELECTRONICS

The vehicle electronics consist of:
» Motor solenoid switch
» DC converter, 24V to 12V
» Switch module
» Charging jack
» Batteries

The electronics system is relatively simple. Two 12V batteries (I used mobility scooter, sealed lead-acid batteries) are wired in series to generate a base voltage of 24V. I store these on the floor, between the bulkhead and the motor, directly under the live shaft. They can be charged without removing them from the vehicle, via a charging jack mounted on the wood frame.

CAUTION: Be careful that the battery electrodes do not contact the live shaft — I learned this the hard way. There are now burnt marks on the shaft where I briefly, directly, shorted the batteries.

The motor runs on 24V and is not currently speed controlled. It is simply turned on or off, and the vehicle's speed is modulated with belt slippage.

There's a switch box located just in front of the seat (Figure H). This box has 2 switches: a 3-position mode switch, and an on/off toggle for the motor. The 3-position switch is used to set the vehicle's electrical system into one of 3 states: All Power Off; Charge Mode (the motor cannot be powered while charging the batteries); and Ready to Run (power to the 24V-to-12V converter and gauges is turned on, and the motor can now be switched on using the toggle).

The converter's primary purpose is to power a continuously operational solenoid that acts as a high-current relay to switch on the motor. The solenoid is powered on its coil side by 12V. (Figure I).

TYING THE MODULES TOGETHER

Final assembly is quite simple: attach the tracks to the frame, install the belts, then attach the body to the frame.

The tracks are attached to the transverse frame sections using 2 bolts per section, for a total of 4 bolts per track. I used 5/16"-diameter hex head bolts 5" long, with 2 flat washers and 1 lock washer on each. Now's the time to align and mount the brake belts in their final position on the frame.

The main drive belt between the motor and live shaft assembly is mounted, then tensioned, by sliding the live shaft assembly in its bearing blocks rearward, away from the motor.

The 2 track drive belts are, by virtue of the way the drive system operates, not under tension, and can therefore be simply placed on the track drive and live shaft pulleys.

The body is attached to the frame via 4 mounting points. Two mounting points are located on the body's bulkhead. A wood-to-machine screw stud (aka hanger bolt) is embedded in each longitudinal frame 2×4 to line up with the mounting holes in the 2×2 cross member on the bulkhead. The hanger bolts protrude internally through the bulkhead; washers and wing nuts are used to secure the body to the frame.

The other 2 mounting points are located toward the front of the vehicle. I made 2 brackets that come off the motor mounts, and welded a nut to the underside of each. Hex bolts (1/4"×2") are used to secure the front of the body to these bracket nuts, through one of the front 2×2 body cross members.

GOING FURTHER

At some point, my son will be too big to ride in the vehicle — so I intend to add servos and a receiver under the seat, and convert the mini tank to R/C! ⊙

UNDERSTANDING THE DRIVETRAIN

During operation, the motor is running constantly at full speed, but the tracks are not engaged until the control handles are pushed forward, engaging the tensioner arm and thereby tensioning the track drive belts (Figure J). To steer, the outside track moves while the inside track remains stopped, forcing the vehicle to pivot around the inside track. To stop, the control handles can be released, resulting in both track-drive belts losing tension; the vehicle will then quickly coast to a stop. To stop faster, both control handles can be pulled rearward, taking the tension off the drive belts and also engaging the brakes.

See video of this drivable mini-tank in action by visiting the full project page: makezine.com/go/mini-tank.

PROJECTS

Remaking History

Chester Rice and the Dynamic Loudspeaker

Written by William Gurstelle • Illustrated by Peter Strain

WILLIAM GURSTELLE is a contributing editor of *Make:* magazine. His latest book, *Defending Your Castle*, is available at all fine bookstores.

Time Required: 1-2 Hours
Cost: $5-$15

Materials

- **Magnet wire, 28 gauge, 60' or 120'** Besides being an electromagnet, the wire coil provides resistance to current flowing through the speaker, which is necessary to protect the circuitry of your amplifier. Most amplifiers are designed to encounter 4 or 8 ohms of impedance from the loudspeaker, and less than that may damage your amp. 28-gauge copper wire is rated at about 0.065 ohms per foot, so 60' will provide 4Ω and 120' will provide 8Ω. If you use different gauge wire, change the length accordingly.
- **Rare earth magnets, ¾" disc, ¼" thick (2)** Make sure you get the normal, *axially* magnetized type — not diametrically magnetized.
- **Cardboard strips, ½"×1½"(2)**
- **Wood board, about 1"×6"×6"**
- **Wood dowel, ¾" diameter, 6" length**
- **Diaphragm, 4" diameter** made of light, stiff material, such as thin balsa wood, a plastic disposable plate, or an empty large-sized tuna fish can. Experiment to see which material provides the best sound.
- **Tissue paper**
- **Epoxy glue, quick setting**
- **Amplifier** compatible with 4Ω or 8Ω speakers
- **Music source** such as a digital music player, smartphone, or phonograph

Tools
- Scissors
- Sandpaper

Build the moving-coil, direct-radiation transducer that has rocked our world for 90 years

CHESTER W. RICE, LIKE HIS FATHER, WAS A GIFTED ENGINEER AND MANAGER FOR THE GENERAL ELECTRIC CORPORATION. In the 1920s, the younger Rice lived in Schenectady, New York, in a leafy, green part of town known as the GE Realty Plot — a cluster of three-story mansions owned by GE executives. Some of these homes had more than six fireplaces, while others had their own ballrooms. Chester Rice's home had something far more practical, considering his job at GE: a fully equipped laboratory.

A large home lab suited the eccentric Rice well. He would start working at 11 or 12 o'clock at night and conduct complex experiments until dawn. In the nighttime peace and quiet of the GE Plot, he could use his ultra-sensitive electronic instruments undisturbed by activity on the street outside.

Chester Rice's accomplishments were voluminous and covered many areas of electrical engineering. He developed short-wave radio technologies and designed one of the first submarine detection systems. He even tested an embryonic radar gun by measuring the speed of trolley cars. But arguably, his most important creation was the first really useful loudspeaker, which he developed with fellow GE engineer Edward W. Kellogg. Music would never be the same.

GETTING AMPED

Prior to Rice and Kellogg's invention,

the only practical way to really listen to a phonograph or a radio was to use headphones. If more than one person wanted to listen, a large acoustic horn was placed over a vibrating diaphragm, but this amplified the sound only minimally. So, in the early 1920s, Rice and Kellogg perfected the *dynamic loudspeaker*, in which an electronic amplifier boosted the signal from a radio or vibrations from a phonograph and then electromagnetically coupled those vibrations to a diaphragm that radiated sound directly. The new system easily filled a whole room with music.

Based on this work, RCA debuted their Radiola 104 "loud speaker" in 1926. It sold for about $250, which is well over $3,000 today. The public thought it well worth the money. Thousands of Radiola 104 speakers were sold.

MAKE YOUR RICE-KELLOGG MOVING COIL LOUDSPEAKER

In this edition of Remaking History, we'll craft a Maker-friendly version of Chester Rice's dynamic loudspeaker (Figure A). The original Rice-Kellogg speaker used two electromagnets — one as the *driver* and one as the *voice coil* — but these days we use cheap permanent magnets for the drivers. Unlike the Radiola 104, our version costs about $15, or less if you happen to have a spool of magnet wire lying around. You'll also need an amplifier and a music source.

1. Wrap one turn of tissue paper around a dowel, then wrap the magnet wire around the paper until you have a coil between ½" and ¾" wide (about 230 turns for 60 feet of wire, or 450 turns for 120 feet). This is your voice coil. Leave 12" of loose wire at the start and finish. Wrap the wire tightly, but not so tightly you can't slide it off the dowel when you're done wrapping.

Gently remove the coil from the tissue paper and dowel, taking care that it doesn't unravel. Smear a thin coat of glue on the coil and let it harden (Figure B).

2. Sand off 1" of insulating paint from the 2 wire ends (Figure C).

3. Glue the voice coil to the center of the diaphragm (Figure D). Let the glue harden.

4. Fold the cardboard strips into Z-shapes as shown in Figure E. Glue the magnets and the Z-shaped cardboard strips to the wood base as shown.

5. Glue the diaphragm to the other ends of the Z-shaped strips as shown in Figure F. Note that the spacing between the voice coil and the top of the magnet affects the quality and volume of the sound. You can experiment to see what height produces the best sound for your coil.

6. Connect the speaker wires to your amp (Figure G) and enjoy your homemade sound!

HOW IT WORKS

How does a loudspeaker make sounds?

Sounds travel through the air as waves. When you strike a drum with a stick, for example, the vibrating drumhead moves the air molecules surrounding it. These air molecules in turn push upon their neighbors, causing them to move also. In this way, a drumbeat moves outward from the drum as a pressure wave in the air, compressing the air molecules before it as it goes. But when the wave passes, the air molecules move back to their previous spacing, temporarily creating low-pressure regions. Sound waves are simply patterns of alternating high-pressure regions called *compressions* and low-pressure regions called *rarefactions* traveling through the air.

When the pressure wave reaches the ear, it pushes the eardrum inward and outward, and the middle ear's nerves convert this vibration to electrical signals that your brain interprets as sound.

Like a drumhead, the vibrating diaphragm of your loudspeaker also causes waves of high and low pressure to travel through the air. But why does the diaphragm move? It's glued to a magnetic coil positioned just in front of a permanent or *field magnet*. When you connect your speaker to a music amplifier, small pulses of electricity, which are shaped by the frequencies of the music, travel through the speaker wires into the coil, turning the coil into a variable *electromagnet* that produces a pulsing magnetic field. As the pulses flow, the electromagnet either attracts or repels the permanent magnet, vibrating the diaphragm to create sound — loud, room-filling, arena-filling sound.

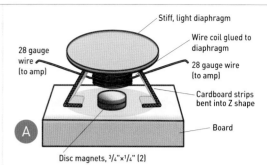

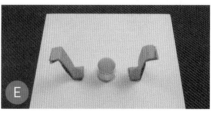

See it work and share your DIY speakers at makezine.com/go/moving-coil-speaker.

PROJECTS
3D-Printed Tourbillon Clock

3D-Printed Tourbillon Clock

Print and assemble this working, large-scale model of a precision watch design

Written by Christoph Laimer

Time Required: A Week
Cost: $40–$60

CHRISTOPH LAIMER was born in Zurich, Switzerland, and grew up nearby. He has a master's degree in electrical engineering from ETH in Zurich, and recently left a career in software to pursue his passion for 3D printed timepieces.

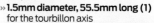

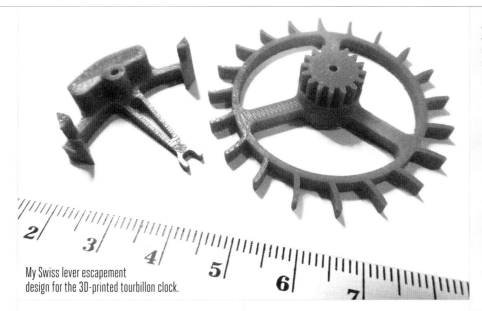

My Swiss lever escapement design for the 3D-printed tourbillon clock.

Materials

- PLA filament
- PETG filament
- Pins, steel or brass:
  - 1.5mm diameter, 55.5mm long (1) for the tourbillon axis
  - 1.5mm dia., 12mm long (1) anchor axis
  - 1.5mm dia., 8.5mm long (1) planet gear axis
  - 2mm dia., 57mm long (3) axis for pinions for minutes and hours wheel
  - 2mm dia., 22mm long (6) axis for basic transmission
  - 2mm dia., 15mm long (1) for attaching mainspring
  - 3mm dia., 22.5mm long (1) axis for mainspring
  - 3mm dia., 31mm long (1) axis for main pinion
- Washers:
  - 3mm (3) for mainspring pinion
  - 2mm (6) for transmission
  - 1.5mm (5) for tourbillon and escapement
- Screws:
  - 1.5mm dia., 5mm long (5) for the going barrel
  - 1.5mm dia., 10mm long (7) 4 for base plate, 3 for tourbillon cage
  - 1.8mm dia., 6.5mm long (5) for ratchet pawls
  - 1.8mm dia., 12mm long (4) for clock face

Tools

- **3D printer** Get the 3D files for free at thingiverse.com/thing:1249221.
- **Drill and screwdriver**

I BOUGHT MY FIRST 3D PRINTER IN 2013 AND IMMEDIATELY STARTED CREATING CUSTOM LEGO GEARS FOR MY KIDS. Next, I challenged myself to design a gear with a Swiss *lever escapement* — the mechanical linkage in a timepiece that swings back and forth and creates the "ticking" sound. This sparked my passion for 3D-printed clock design, and my first 3DP wall clock was ticking 6 months later.

Then I got an Ultimaker 2 and a new challenge. I was obsessed with Vianney Halter's Deep Space Tourbillon, a *Star Trek*-inspired wristwatch whose mechanics rotate visibly at its center. In watchmaking, the *tourbillon* is a slowly spinning cage for the watch's escapement and balance wheel, which is meant to average out the effects of gravity on the timepiece's accuracy. With advancements in modern watchmaking a tourbillon is unnecessary, but designers still include it as a demonstration of their skills. It's a piece to really show off, and it takes 100% focus to achieve.

With this in mind, I knew my 3D-printable watch needed to have a tourbillon. Like the Deep Space Tourbillon, the ticking-unit should be in the center of the watch, and the hands should rotate around it, using big internal gears. The result is a reinvention of a classical tourbillon, adapted to 3D printing.

This windable watch consists of 51 printed parts, 15 pins, 14 washers, and 21 screws. When it's all screwed and snapped together, it's clock-sized: 4" (102mm) in diameter. I've shared the 3D files on Thingiverse so that anyone can make their own.

DESIGNING A TOURBILLON

My first experiments in printing a tourbillon involved the creation of small-module 3D-printed gears. For gears, a *module* is a unit that indicates the distance between gear teeth. It's the ratio of the reference diameter of the gear divided by the number of teeth. It turned out that my Ultimaker could print working gears with module 0.5 — half the size of my Lego gears. I decided to design the gears with module 0.7 to be on the safe side.

Next, I needed to design a printable escapement small enough to fit inside the tourbillon cage. By arranging the balance wheel and the escapement gear in a concentric, coaxial orientation, I minimized the build volume. For driving the escapement, a smaller gear was added to the design.

The transmission in the tourbillon is similar to a planetary gear with a stationary annular gear, and the tourbillon cage as carrier. Designing this in Autodesk Fusion 360 was surprisingly easy, and didn't require much prototyping. The tourbillon turns at 1 revolution per minute — so the second hand is attached directly to it.

The design of the minutes and hours gears and hands was straightforward. I started with the design of the clock face, and arranged the gears behind the tourbillon. The gear ratio is simple mathematics.

A bigger challenge was the design of the mainspring. I designed and 3D printed a couple of spiral-shaped objects, and tortured them until they broke. I also

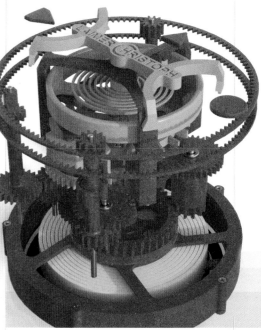

An exploded view of what the tourbillon clock looks like without its casing.

PROJECTS | 3D-Printed Tourbillon Clock

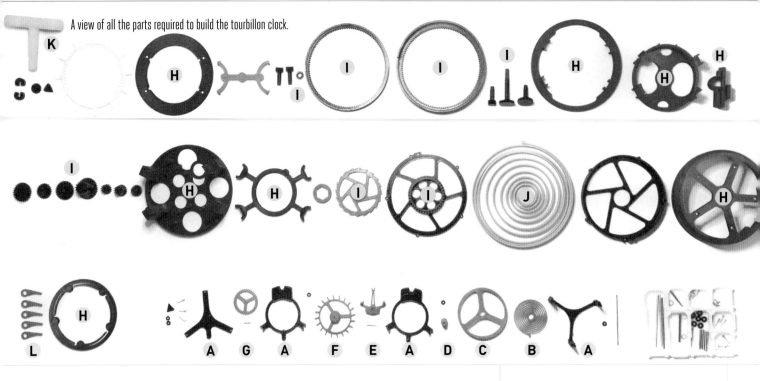

A view of all the parts required to build the tourbillon clock.

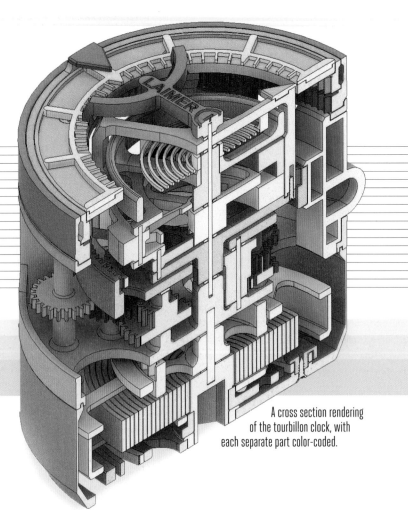

A cross section rendering of the tourbillon clock, with each separate part color-coded.

learned (from Google) that a relaxed mainspring must have a special shape in order to keep the driving force constant. From my experiments, I learned that PLA wasn't suitable — it deforms relatively quickly, and the driving force diminishes over time as the spring wears out. The behavior with PETG filament is still not the best, but compared with PLA it's much better. In the future I'd like to experiment with other materials. The final 3D-printable mainspring has an unwound length of about 2 meters, and takes a long time to print.

I started printing my first tourbillon watch in June 2015, and the final watch was working by December. When I decided to publish the tourbillon design on Thingiverse and YouTube, I never expected such a response. Now I'm receiving offers to work on other 3D-printable designs. I'm hoping to start a new career, while continuing my passion for watch design as a side project.

BUILD YOUR OWN TOURBILLON CLOCK

Building this project requires patience and quality control. You're building a watch mechanism much larger than professional watchmakers have to fuss with, but still, developing your eye for print quality, weight, and material strength is key to the project's success and the accuracy of your timepiece.

Christoph Laimer

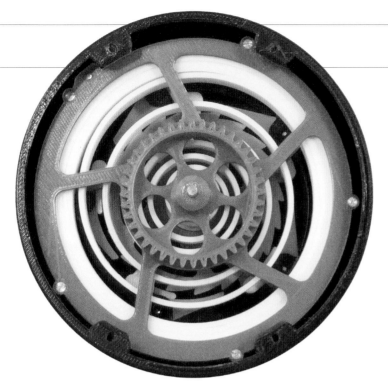

For the mainspring "going barrel" assembly, align the barrel as shown and slide it into the case. The pawl unlock key (included as a 3D-printable part) is a useful tool to keep the pawls in place during this step.

Print all the pieces, noting the guidelines below. Then build the 2 main modules — the tourbillon and the mainspring "going barrel" — and assemble the clock gearing by following the video build guides online at makezine.com/go/3d-tourbillon.

Everything inside the **tourbillon cage** A — i.e. the **hairspring** B, **balance wheel** C and **pin** D, **anchor (lever)** E, **escapement wheel** F, and **planet gear** G — was printed in PLA at high resolution (0.06mm layer, 0.8mm shell). All other parts were printed at normal resolution (0.1mm layer, 0.8mm shell). All parts can be printed without support, except the pawl key (not shown here). I got best results using my Ultimaker 2 with a 0.4mm nozzle.

The infill of the anchor is 80% in order to create a more balanced center of gravity. The rest of the pieces have 30% infill.

For the **case** H I used PETG (slightly bendable, shock absorbing). All **gears** I are printed with PLA (harder and less friction). In these photos, black and yellow parts are PETG, orange and red are PLA.

The **mainspring** J is printed in PETG (PLA would probably break after a while). I switched off the "combing" setting in Cura. While this is a cool feature for ordinary parts, it causes problems with large spiral-shaped parts, as the print head does many useless traveling moves. The nozzle oozes during these moves, and when it resumes printing it can be empty of material; the resulting under-extrusion can be disastrous.

The hairspring needs to be printed in PLA. Other materials can work, but you'll find that the watch runs too slow, or too fast. Because of this, any material substitution requires a new spring design.

Small holes usually don't print very accurately, so use a drill to smooth the inner surface — the balance wheel especially needs to rotate with very little friction and very little play. If you don't find pins or screws with the recommended diameter, you can use slightly larger fasteners — there's some room to drill the holes out.

For best adhesion, print on a heated glass bed cleaned with a mix of alcohol and water. To remove parts, pour a few drops between the part and the glass. The effect is miraculous — the parts can be removed immediately without applying any force.

Besides the **winding key** K there's also a pawl key to unlock the **ratchet pawls** L so that you can fully unwind and relax the mainspring when your watch is not in use. This certainly will extend its lifetime.

You can even print a chain for pulling it out of your pocket. OK, it needs a big pocket!

See the tourbillon clock in action and build yours at makezine.com/go/3d-tourbillon.

3 Fun Things
to 3D Print

MARBLE MACHINE #3
By Tulio Laanen
thingiverse.com/thing:1385312

This is a relatively large design that takes about 10 hours to print but has the advantage of requiring no supports. Once printed, the 4-piece assembly is simple and the payoff is very fun. Spinning a knob on the top of the machine rotates a worm gear that carries marbles from a reservoir at the bottom to the top, where they cascade down a series of twisting chutes.

SD CARD MOUNTAIN
By 3D Brooklyn
thingiverse.com/thing:1362048

There's no shortage of SD card holder designs available online, but you'll be hard pressed to find one as cleverly designed as this. The microscopic mountain fits five cards at an angle, adding some texture to the rocky range. The print is all one piece, with no need for support.

SPIROGRAPH
By Valdis Torms
thingiverse.com/thing:905849

The low-profile design of these Spirograph-style drawing gears makes for a super-quick print with a fun payoff as soon as you pry it from the print bed. We found it useful to enlarge the holes with an awl after printing to save a few trips to the pencil sharpener.

PROJECTS | Percusso Drumbot

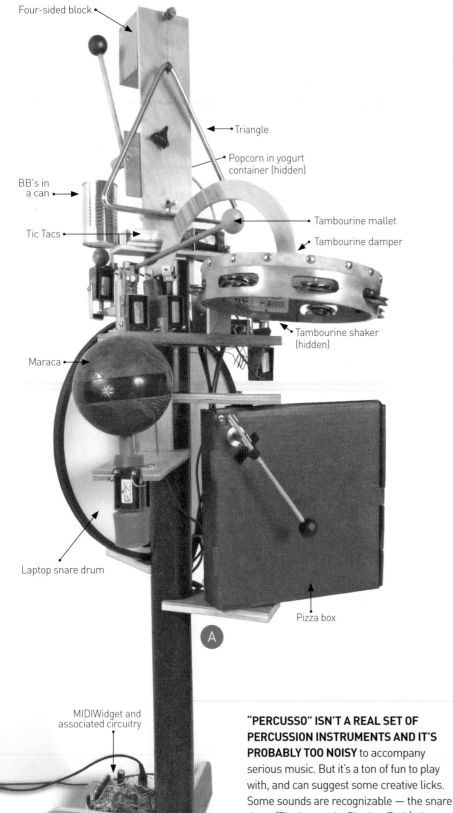

How I built a crazy percussion robot with simple solenoids and a MIDIWidget

Percusso: A MIDI-Controlled Percussion Bot

Written by Larry Cotton

"PERCUSSO" ISN'T A REAL SET OF PERCUSSION INSTRUMENTS AND IT'S PROBABLY TOO NOISY to accompany serious music. But it's a ton of fun to play with, and can suggest some creative licks. Some sounds are recognizable — the snare drum (The Laptop by Rhythm Tech), the maraca, a triangle, and a shaken tambourine — but that's about it. The other "instruments" — BBs in a tomato sauce can, a 4-sided block, popcorn in a Chobani yogurt container, a mallet and damper for the tambourine, Tic Tacs, and my editor's favorite, a red mallet hammering on a 10" pizza box — mostly just make interesting noises (Figure A).

They're all computer-controlled by my Yamaha P-105 digital piano via MIDI. I've temporarily dedicated 12 high notes of that keyboard to Percusso. MIDI, if you don't know by now, is how computers and (typically) musical devices communicate. With the proper hardware and software, any key on a digital keyboard can be "mapped" to play another electronic musical or percussive instrument — or most any electromechanical thing.

MIDI signals can control solenoids, which is what I've done here, or lights, valves, relays, electro-pneumatic devices, motors (stepper and ordinary), and more. Scary Halloween scenes and Christmas light extravaganzas are often controlled by MIDI. (More about MIDI later.)

BUILDING PERCUSSO

I chose my instruments largely by what I had on hand. I did buy a triangle because Percusso seemed to need something tinkly to go with the other sounds.

After much trial-and-mostly-error experimentation, I settled on the noisemakers and sketched a tentative physical arrangement. Since my music studio is pretty full of, um, stuff already, I decided on a vertical layout, with the smallest instruments on top and the larger ones near the bottom. The whole shebang doesn't take up very much floor space and can be moved relatively easily. I used a 4-foot piece of 2"-diameter PVC pipe for the main support structure, cutting ½"-wide slots in it to hold the percussion instruments. Two half-lap-jointed pieces of ½" plywood serve as legs.

For assembly, portability, and troubleshooting, I wanted the instruments to be self-contained, easily removable modules. Each module includes a sound-creator, its solenoid, the mechanical strikers and pivots, an RCA jack, and a mounting surface, generally ½" plywood (Figures B, C, D, and E, following page).

Since the different instruments need to be shaken or struck in different ways, I improvised several strikers, brackets, and linkages to the solenoids. These I made primarily from standard sizes of aluminum and wood.

The most challenging mechanism was the maraca shaker. Shaking a maraca by hand requires a bit of wrist motion to sync the beans' movements in the shell. Human wrists turn out to be rather tricky to emulate with a solenoid, so I added a spring and some padding here and there. The solenoid actually lifts the maraca's shell, and its

LARRY COTTON is a semi-retired power tool designer and math teacher. He loves music, computers, birds, photography, electronics, furniture design, and his wife — not necessarily in that order.

Time Required: Several Weekends
Cost: $200 and Up

Materials

- » **"Instruments" of your choice** from tambourines to pizza boxes
- » **Solenoids, pull type, 12V–24V DC** Try surplus at All Electronics, Alltronics, and American Science and Surplus. Selection varies; I found some without plungers, but 5/16" bolts worked fine as DIY plungers.
- » **Compression springs, various** with inside diameters to fit over solenoid plungers
- » **MIDIWidget interface board** about $60 at midiwidget.com
- » **MIDI keyboard**
- » **Computer** with MIDI sequencing software and/or DAW software (see article)
- » **Power supplies, switching, 12V and 24V DC**
- » **Darlington transistors** ULN2803A array chip (recommended) or TIP120 type
- » **Wood, various sizes, flat and linear**
- » **Plywood, ¼" and ½",** good quality
- » **Wood dowels, 1/16"–½" diameters**
- » **Wood balls, 1" diameter** for strikers
- » **Aluminum extrusions** in common sizes
- » **Plastic sheet scraps**
- » **PVC pipe, 2" diameter**
- » **Fasteners of all types** especially various sheet metal and machine screws and nuts
- » **Glues and tapes**
- » **Wire, 22 gauge, insulated**
- » **RCA audio plugs, sockets, and cable**
- » **Brazing rod, 3/32" diameter** makes great pivots and travel limiters
- » **Incidental noise deadeners** such as Creatology foam sheets
- » **Spray paint** for PVC pipe
- » **Lacquer, Deft satin finish** for wood parts

Tools
Some of my more useful tools are:
- » **Shopsmith** used mostly as a sander
- » **Band saw**
- » **Drill press with ½" chuck**
- » **Table saw**
- » **Cordless drill, 3/8" chuck, variable speed**
- » **Dremel high-speed rotary tool** and lots of accessories
- » **Jigsaw and hacksaw**
- » **Drill bits** from 1/16" to ½"
- » **Spade or brad-point bits** from 3/8" to ¾"
- » **Hammer, pliers, and screwdrivers**
- » **Soldering gun or iron**
- » **Threading taps: 6-32 and 8-32**
- » **Files**
- » **Sandpaper**

USING SOLENOIDS FOR PERCUSSION
TIPS, TOOLS, AND MATERIALS

1. Try to make the solenoid plungers (cores), and whatever is connected to them, move with as little friction as possible and with enough travel to hit or shake something. Keep parts in alignment. Don't depend on the solenoid for too much strength or stroke.

2. When a solenoid plunger returns after making a percussive strike, you may not want to hear its return noise. To keep these incidental noises to a minimum, plan to use lots of foam between and around parts that hit each other. Use springs that are easily compressible, yet quickly return the plungers to their home positions.

3. The solenoids I bought are "pull" type, but the "push" end of the solenoid is usable too by enlarging a small hole in the other end of the housing.

4. Different sound modules may need different voltages. 12V DC seems adequate to do simple horizontal movements, but you may need 18V to 24V to do vertical, elaborate, longer, or stronger movements.

5. Ensure your power supplies can supply adequate current to the solenoids. If you plan to drive several simultaneously, the current is additive. I find playing no more than 2 instruments simultaneously to be satisfactory.

6. Connect MIDIWidget and the noise modules with RCA plugs, sockets, and cable.

7. Add diodes across the solenoids to minimize transient voltage spikes. Watch their polarity.

PROJECTS | Percusso Drumbot

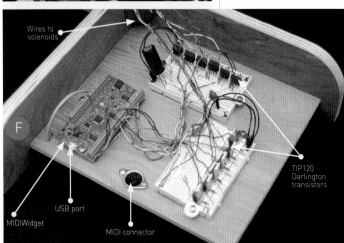

handle pivots at its far end, so 24V DC were necessary. The keyboard should be played in sync with the natural rattle of the beans.

INTERFACE BOARD
My computer and solenoids are interfaced with John Staskevich's MIDIWidget (Figure F). It's widely available now, but I got mine as an early reward from his Kickstarter campaign. John has mass-produced many similar products, such as the MD24 decoders I used for my MIDI marimba (youtu.be/UQ35iQFeock). Check out some of his other activities at highlyliquid.com and codeandcopper.com.

MIDIWidget, which can control up to 24 devices, is quite easy to use and eliminates the arcane programming associated with earlier interfaces. I connected MIDIWidget to my computer, launched its software configurator (where the keyboard notes are mapped to the solenoids), and connected my solenoid control circuits to MIDIWidget. I elected to pay a bit more for the screw terminals (I used half of them) and highly recommend this. The board receives power either via USB or a 5V DC regulated supply and comes with a standard MIDI connector.

The MIDIWidget's signals are tiny 5V pulses not strong enough to directly drive solenoids or anything that requires significant current, such as motors or relays. John recommends using a Darlington transistor array chip ULN2803A between MIDIWidget and solenoids, but I had a bunch of TIP120 Darlington transistors, so I used them instead. Bonus: Each TIP120 can handle more current than a single output of the ULN2803A. When MIDIWidget decodes a MIDI stream, the Darlingtons amplify the 5V pulses to drive the higher-current devices. Figure G shows the circuitry necessary for each note.

My solenoids run on 12V–24V DC and draw a fair amount of current, so I used a couple of beefy power supplies (12V and 24V DC) I had on hand. Most wall-warts won't be up to this task unless they're switching power supplies with higher rated loads.

SOFTWARE
A computer records what's being played on a digital keyboard using MIDI sequencing software. MIDI sequencers can handle far more than the 12 tracks I can record for Percusso. Fortunately, you don't need to know how to play a keyboard: Think of the keys as just a bunch of sophisticated switches that turn stuff on and off.

You can play and edit a MIDI "performance" with basic software. To include audio (.wav or .mp3 files), you'll need Digital Audio Workstation (DAW) software, which is more sophisticated. With a DAW, your audio and MIDI tracks (sequences of signals) can be synced to each other.

I use an ancient DAW — Home Studio 2002 by Cakewalk — which runs fine under Windows 7. Cakewalk's newer offering is SONAR. Other people like Steinberg's Cubase MIDI sequencing software. Anvil Studio offers a free sequencer that can record unlimited MIDI tracks and two 1-minute audio tracks. For another $20, you can add up to 8 audio tracks.

Should you decide to build your own version of Percusso, I urge you to experiment, be creative, and plan to make lots of mistakes. Most of all, have fun! ✱

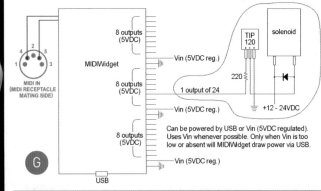

Watch video of Percusso in action and share your music bot ideas at makezine.com/go/percusso.

Ponytrap:
A Robot Drummer with Arduino

Written by Quentin Thomas-Oliver

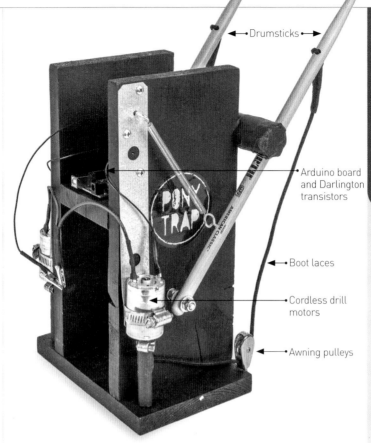

- Drumsticks
- Arduino board and Darlington transistors
- Boot laces
- Cordless drill motors
- Awning pulleys

QUENTIN THOMAS-OLIVER made up Ponytrap with his wife, Hillary Thomas-Oliver. They like to think of their art as tribal music for modern primitives: industrial music on classical instruments ... along with a couple of robots.

Time Required: 2–3 Hours
Cost: $75–$80

Materials
- **Cordless drill motors, 12V DC, with pinion gear (2)** We used Mabuchi RS-545SH.
- **Drumsticks (2)** We used Vic Firth 7A's because we like the response of the lighter stick.
- **Arduino Uno microcontroller board** with USB cable
- **Boot laces**
- **Pine board, 1×6, 4' total length**
- **Deck screws, 1½" (12)**
- **Wood screws, #8, ½" (6)**
- **Hex bolts, ¼", 2" long (2)**
- **Lock nuts, ¼", with washers (4)**
- **Threaded eyelets or #12 screw eyes (4)**
- **Steel strap ties, 1¼"×9" (2)** Simpson #LSTA9
- **Hose clamps: #4 (2) and #20 (2)**
- **PVC pipe, ¼", 5" total length** for drive shafts
- **PVC pipe, ½", 8" total length** for spindles
- **Awning pulleys, ¾" (2)** fast eye type
- **Extension springs**
- **Lantern batteries, 6V (2)**
- **Female push connectors (4)** small gauge
- **Breadboard, small**
- **Hookup wire**
- **Transistors, TIP120 (2)**
- **Diodes, N4004 (2)**
- **Pipe insulation, scraps** for the stick return "bumper"

Tools
- **Hacksaw**
- **Drill with bits: ¼" and ³⁄₃₂"**
- **Screwdrivers, slotted and Phillips head**
- **Adjustable wrench**
- **Pliers with wire cutters**
- **Pencil**
- **Ruler**
- **Gaffer's tape**

Build a hard-charging drumbot with real drumsticks and your favorite microcontroller

A COUPLE OF YEARS AGO, I WAS PLAYING MUSIC AND THE IDEA TO BUILD MY OWN DRUMMER JUST KIND OF MATERIALIZED. After dozens of burned-out motors and a lot of trial and error, my band Ponytrap now has a solid drumline made entirely of robots.

A big part of our aesthetic was to make the drum both physically and visually satisfying. It may be simpler to use smaller, faster actuators, but we really like to see, hear, and feel the drums beating away! Our big machines are fun to watch — and they pound as hard as a human drummer.

USING DC MOTORS FOR PERCUSSION
This drumming robot uses two drumsticks controlled by an Arduino microcontroller, and all the beats are written directly into the Arduino software. For strength and power, we use two 12V DC, 24,000rpm cordless drill motors, one for each stick. Each motor spins a drive shaft made of ¼" PVC pipe connected to a spindle of ½" PVC, which rapidly reels in a string that draws the drumstick down to the drumhead for a strike. After the strike, a return spring draws the drumstick back up to prepare for another strike. The motors are turned on and off by the Arduino via TIP120 transistors and powered by 6V lantern batteries wired in series to 12V (Figure Ⓐ).

We use this machine with a snare, but any drum is suitable. (If you don't have drums handy, plastic buckets make excellent substitutes.) You can get all the parts from retail stores and build the entire thing with simple hand tools. At the end of the day, you'll have an Arduino-controlled drumming robot that can play complex beats of various speeds — and will always show up to rehearsal. ●

See the complete build instructions, video, and code at makezine.com/projects/make-robotic-drum-using-arduino-uno.

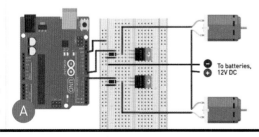

To batteries, 12V DC

PROJECTS | Macramé Pixel Art

Mario Play Cubes

Written by Jane Stewart

JANE STEWART is a big crafting fan who lives in the U.K. As a kid she'd spend every holiday doing a project; at 11 she received a friendship braiding kit for Christmas, setting her on the macramé path.

Transform vintage video game pixels into an easy macramé pattern — no knitting needed!

THIS IS A SOFT PLAY CUBE THAT I MADE WITH MY 5-YEAR-OLD NEPHEW IN MIND, so he can have a real-life squishy *Super Mario World* in his room!

I've done macramé for 20 years. I learned as a schoolgirl when I started a craze for friendship braiding, and I worked out how to braid straight lines and swap over colors while experimenting.

You can easily convert simple pixel graphics, like the sprites in vintage video games, to a macramé pattern for making these fun soft toys. And you can scale them up to pillow size or bigger if you can find yarn bulky enough!

PIXELS TO KNOTS

I created these graphics of Mario and various power-ups using the classic Super Nintendo game *Mario Paint*, but you can use any drawing program, or just work out your grids on paper.

I adapted the 16×16 pixel grids first worked out by Nintendo (Figure Ⓐ) to fit in my macramé panels. Because the double half hitch knots are slightly higher than they are wide, I had to squash the images a little to keep things square (or cubic). So I simply deleted the top and bottom rows from my *Mario Paint* grids to create these 16×14 pixel instructions.

1. LEARN THE KNOT

Figure Ⓑ shows how to do the basic double half hitch knots I used; I also recommend you check out video tutorials on YouTube. I'm right-handed, but I find it easier to hold the structural or "foundation" yarn taut with my right hand while doing the fiddly knotting with my left. You can do it the other way around if you find this more comfortable.

2. KNOT SOME PLAIN ROWS

For the structural yarn, cut 48 threads of black yarn, each 28" (70cm) long. Two black threads will end up inside every macramé knot. Each panel consists of 3 rows of a

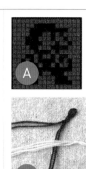

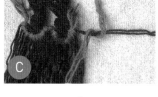

Time Required:
8–10 Hours

Cost:
$10–$30

Materials
- **Yarn, 100g balls, lightweight** aka double knit (DK) weight. Use as many colors as you like. I used 7 in this project.
- **Stuffing** such as polyester fiberfill, fabric scraps, packing foam, dry beans, or pasta

Tools
- Scissors
- **Computer with drawing program (optional)** Pencil and paper works too.

dominant color (for example red on the Mario panel), then 14 rows where the colors are swapped over to make the picture, then 3 more dominant color rows at the bottom. Each panels is 24 knots long and 20 high.

Figure **C** shows the first row of red knots completed. The structural yarn is tied temporarily in 2 clumps of 24 threads to stop them from shifting about.

3. KNOT SOME PICTURE ROWS
Each pixel stands for a double half hitch knot over the structural wool. Leave 3 plain rows above and below the picture, and 4 knots on either side of it.

Figure **D** shows the first picture row being made by swapping between the red background color and the black color of the Mario pixels. Tuck the ends of the black yarn behind the panel as you finish each row. These can be tied together at the back of the panel when it's complete.

Once 3 rows of red are completed you can undo the big knots in the structural wool and begin the Coin panel, starting at the first row of red and moving away from the Mario.

Four complete picture panels fit on the 28" black structural yarn (Figure **E**).

4. MAKE THE 2 SMALLER PANELS
I made the Question Block panel in the same way on 8" (20cm) lengths of yellow structural yarn, and the Star on 8" blue structural yarn.

5. ASSEMBLE THE CUBE
The cube's sides are joined by simply knotting end of row to end of row. First tie the 2 small panels on either side of Mario to make a cross shape (Figures **F** and **G**).

Then tie the cube together at the corners, inside out, leaving the Fire Flower panel loose.

6. STUFF AND CLOSE UP
Turn the cube right side out. Fill it up with stuffing, then join the sides of the Fire Flower panel to the rest of the cube (Figure **H**), and tuck in the ends. You're done!

7. EXPERIMENT
I've used various different stuffings in my play cubes, including packing foam, plastic bags, leftover fabric, and even dry pasta for a noisy play cube.

You can make these as big as you like or as tiny as you can manage. Use thin wool or very thick, it's up to you. Make your knots on a single thread for tiny cubes, on 2 threads as I've done here, or on 4 threads for big cubes. I've made cubes in different sizes, and cuboids as well — all the better to build Mario forts with.

Get complete step-by-step photos at makezine.com/go/mario-macrame. Learn more about getting started in macramé at makezine.com/go/macrame-101.

PROJECTS

Book Excerpt — *Make:* Bluetooth

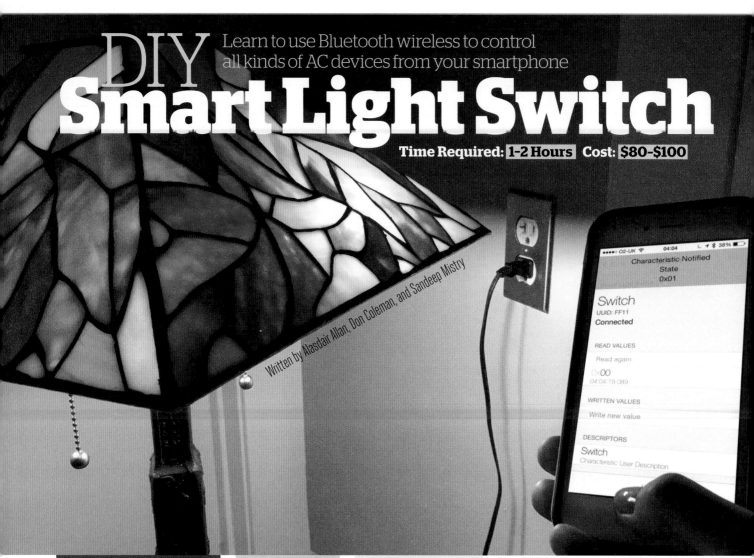

DIY Smart Light Switch

Learn to use Bluetooth wireless to control all kinds of AC devices from your smartphone

Time Required: 1-2 Hours   Cost: $80-$100

Written by Alasdair Allan, Don Coleman, and Sandeep Mistry

ALASDAIR ALLAN is a scientist, author, hacker, tinkerer, and *Make:* contributing editor (makezine.com/author/aallan) who spends a lot of time thinking about the Internet of Things.

DON COLEMAN is a lifelong engineer and seasoned PhoneGap developer.

SANDEEP MISTRY is a software engineer who has created several open source BLE libraries.

This project is adapted from the book *Make: Bluetooth* by Alasdair Allan, Don Coleman, and Sandeep Mistry, available at the Maker Shed and at fine bookstores.

Materials
» Arduino Uno microcontroller board
» Adafruit nRF8001 Bluefruit LE Breakout board Adafruit #1697, adafruit.com
» Solderless breadboard
» Jumper wires
» LED (1)
» Resistors: 220Ω (1) and 10kΩ (1)
» Tactile switch, momentary pushbutton for prototyping
» PowerSwitch Tail for controlling real light bulbs or other AC devices

Tools
» Computer with Arduino IDE software free at arduino.cc/downloads
» Soldering iron
» Small screwdriver

ONE BIG PROBLEM WITH SMART LIGHT BULBS IS THAT THE SMARTS ARE IN THE BULB. They don't play nice with the light switch on the wall. In fact, you can make most smart light bulb systems unresponsive just by using your wall switch! We really need to replace the switch, not the bulb.

WHAT IS A SMART SWITCH?
A smart light switch not only lets you turn the light on and off using the switch itself, but also works remotely via Bluetooth LE wireless networking, so you can control your lights using your smartphone or other mobile device. The switch should also know its current status — in other words, whether the bulb is on or off — and send a notification over Bluetooth to subscribed apps whenever the switch is toggled, so they can update their local status too.

78   makershed.com

Here's how you can connect and configure Bluetooth control of a lamp or, really, any AC-powered device in your home or workshop, using an Arduino microcontroller as the smarts.

1. BLINK AN LED

Blinking an LED is the "Hello, World" of hardware — an easy test to ensure that your Arduino is set up correctly. Grab your Arduino board, a breadboard, an LED, a 220Ω resistor, and some jumper wires, and follow Massimo Banzi's tutorial at makezine.com/projects/make-an-led-blink-with-your-arduino. The example code will set the Arduino's pin 13 as HIGH for 1 second (switching the LED on) and then LOW for 1 second (off), and so on.

2. ADD A PUSHBUTTON SWITCH

Now connect the little tactile pushbutton and a 10kΩ resistor as shown in Figure Ⓐ.

> **NOTE:** What we've done here with the resistor is called debouncing the button. If you simply connected one side of the button to +5V and the other to pin 4 of the Arduino, then pin 4 would be floating because it's not connected to anything. Then our code would work unreliably; sometimes it would detect button pushes that don't exist. To avoid that, you're using the resistor to "pull down" the pin connection to GND. Now the pin is no longer floating; instead it is pulled to a LOW state.

Download this project's Arduino code from github.com/MakeBluetooth. In the Arduino IDE, open the Library Manager and type **BLEPeripheral** into the search window, then select the library *BLEPeripheral.h* and click Install.

Next open the sketch *ble-smart-switch.ino* in the Arduino IDE, and upload it to your Arduino. Now your LED should turn on when you push the button, and off when you push the button again.

3. ADD THE BLUETOOTH MODULE

Now that you have a working pushbutton light switch, let's add Bluetooth LE. The Bluetooth board we're using is the Adafruit Bluefruit LE board based around the Nordic Semiconductor nRF8001 chipset. Wire the board to the Arduino as shown in Figure Ⓑ, using pin 2 for RDY and pins 9 and 10 for RST and REQ, respectively.

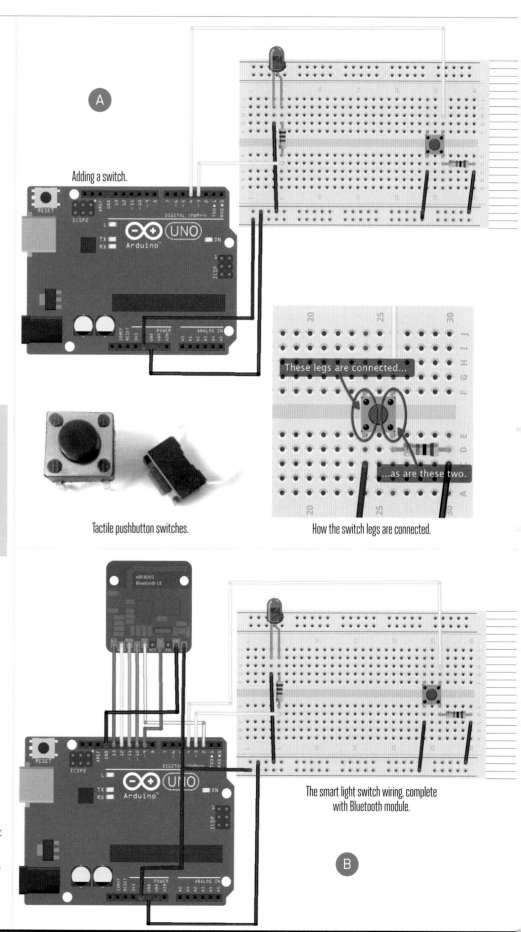

Adding a switch.

Tactile pushbutton switches.

How the switch legs are connected.

The smart light switch wiring, complete with Bluetooth module.

PROJECTS

Book Excerpt — *Make: Bluetooth*

A BIT ABOUT BLUETOOTH LE

The Bluetooth Low Energy wireless protocol (BLE, also marketed as "Bluetooth Smart") divides the world into *peripheral devices*, like sensors and speakers, and *central devices*, like your smartphone or laptop. Peripherals can either connect directly to a central device, or broadcast data to any device in range by sending out "advertising packets." Once connected, the central device can get a list of *services* offered by the peripheral.

So, for our Smart Light Switch, we're going to create a BLE peripheral with a single service. Our service has two *characteristics*: a readable/writeable characteristic called "Switch," which we'll use to flip the switch on and off, and a second characteristic called "State," which can notify us of changes in the status of the switch. Luckily every service and characteristic has its own Universally Unique Identifier (UUID) number, so we can easily plug existing BLE services right into the code for our project.

Our service will therefore look something like this:

LIGHT SWITCH SERVICE FF10

Characteristic	UUID	Properties	Comment
Switch	FF11	read,write	1 on, 0 off
State	FF12	notify	1 on, 0 off

Optionally, we can also use *descriptors* — here we'll use the Characteristic User Description (UUID 0x2901) — to provide a text description of our service to an end user.

4. PROGRAM YOUR SMART SWITCH

Open the Arduino sketch *ble-light-with-powertail.ino* and read along with the comments to see how your Smart Light Switch code will establish the service:

» Create a peripheral instance using the BLEPeripheral library.
» Create a `lightswitch` service with UUID of `0xFF10`.
» Create the Switch and State characteristics and descriptors.

```
BLEPeripheral blePeripheral = BLEPeripheral
(BLE_REQ, BLE_RDY, BLE_RST);
BLEService lightswitch = BLEService("FF10");
BLECharCharacteristic switchCharacteristic
= BLECharCharacteristic("FF11",
BLERead | BLEWrite);
BLEDescriptor switchDescriptor =
BLEDescriptor("2901", "Switch");
BLECharCharacteristic stateCharacteristic =
BLECharCharacteristic("FF12", BLENotify);
BLEDescriptor stateDescriptor =
BLEDescriptor("2901", "State");
```

And then configure it:
» Set the Local Name (for generic Bluetooth access) and Device Name (for broadcast in the peripheral's advertising packet).
» Add your service characteristics and descriptors as `Attributes` of your peripherals instance.
» Advertise the Bluetooth LE service, and poll for Bluetooth LE messages.
» Set both the switch and state to be on if the button is pushed, off if it's released.

```
pinMode(LED_PIN, OUTPUT);
pinMode(BUTTON_PIN, INPUT);
blePeripheral.setLocalName("Light Switch");
blePeripheral.setDeviceName("Smart
Light Switch");
blePeripheral.setAdvertisedServi
ceUuid(lightswitch.uuid());
blePeripheral.addAttribute(lightswitch);
blePeripheral.addAttribute(s
witchCharacteristic);
blePeripheral.
addAttribute(switchDescrip tor);
blePeripheral.addAttribute(
stateCharacteristic);
blePeripheral.addAttribute(stateDescriptor);
blePeripheral.begin();
```

Save and upload the sketch to the board. Now you can turn the LED on and off as normal using the button, and if you open the Serial Console you'll see the string `Smart Light Switch` appear, with further messages every time you push the button to turn the LED on or off.

In addition, now you can "throw" the switch using Bluetooth LE, because we've assigned an event handler to be called when a write command is made on the peripheral: `switchCharacteristic.setEventHandler(BLEWritten, switchCharacteristicWritten);`

And at the bottom of the sketch, after the `loop( )` function, we've added the handler function itself. Now you can control the LED via Bluetooth LE!

5. TEST YOUR BLUETOOTH SERVICE

All you need now is a generic Bluetooth LE explorer app so you can examine and trigger your service. Go to your preferred app store and install LightBlue (iOS) or nRF Master Control Panel (Android) on your smartphone or tablet. The two apps present the same information in slightly different ways.

Opening either app will start it scanning for Bluetooth LE devices. You can choose a peripheral from a list of nearby devices and explore information about that connected peripheral, its services, and characteristics.

Exploring your Smart Light Switch in the LightBlue app.

Now take a look at your Smart Light Switch in LightBlue (Figure C). Tapping through from the Smart Light Switch in the peripherals list, you can see the advertisement data for the service showing our two characteristics: Switch, which is Read/Write, and State, which is Notify. You can register the LightBlue app for notifications when the LED state changes by tapping on "Listen for notifications" in the State characteristic screen.

The Switch characteristic screen shows the current value of this characteristic, which should be **0x00**, meaning the LED is off. Tap on "Write new value" to open the editor. Enter **01** and hit Done; the LED should turn on and the screen should show the new value as **0x01**. If you registered for notifications, you should also see a drop-down to tell you the value has changed (Figure D).

If you have the Serial Console open you should also see the message **Characteristic event: light on** printed in the console. Finally, if you push the tactile button, you should see a further notification in LightBlue that the LED state has changed back to **0x00**.

That's it — you've created a working smart light switch!

6. CONNECT A REAL LIGHT BULB

Now connect your smart switch to an actual lamp. The PowerSwitch Tail (Figure E) simplifies our lives by hiding all that nasty AC electricity and letting us use a relay and our Arduino board to switch real mains-powered devices on and off. Nice.

Connect 3 wires to the PowerSwitch Tail's screw terminals: the left terminal (labeled +in) is for +5V; the middle (labeled -in) is the signal wire; and the right is Ground. Then wire up the Arduino, switch, and PowerSwitch Tail as shown in Figure F.

Plug the PowerSwitch Tail into the wall, and then plug a mains-powered lamp or other electrical device (maximum draw 15A at 120V) into the PowerSwitch Tail socket.

The PowerSwitch Tail can be wired either as "normally open" or "normally closed." From a safety perspective, it makes sense to use the normally open configuration here: Power will only flow while the signal wire from the Arduino is pulled LOW, otherwise the lamp remains "off."

Since pulling the signal wire LOW rather than HIGH is what triggers the relay, we have to flip the logic for the **LED_PIN**. Go back into the code and you'll see that everywhere there was a `digitalWrite(LED_PIN, HIGH);` we have changed it to `digitalWrite(LED_PIN, LOW);` and vice versa.

Now, instead of controlling an LED, you're controlling a real lamp with your phone!

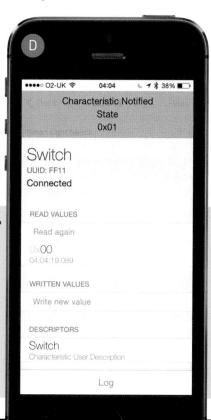

You're notified that the state of the LED has changed.

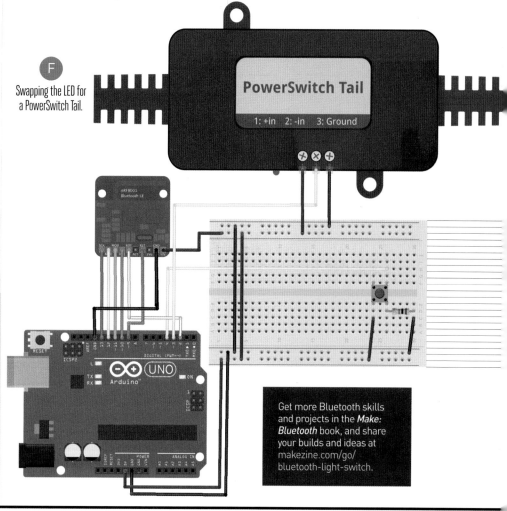

F. Swapping the LED for a PowerSwitch Tail.

Get more Bluetooth skills and projects in the *Make: Bluetooth* book, and share your builds and ideas at makezine.com/go/bluetooth-light-switch.

PROJECTS

Giant Vortex Air Cannon

Written by Tom Heck

Bring the boom with supersized smoke rings that go the distance

Time Required: 10-12 Hours **Cost:** $100-$200

TOM HECK is a dad, Maker, and banjoist. Most of his projects don't fit in his car very easily. Tom is passionate about the Maker Education Movement and is the VP for education initiatives at Makey Makey.

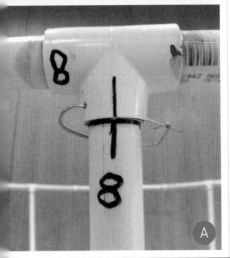

I BUILT THIS GIANT VORTEX AIR CANNON to "bring the boom" to Maker Faire Atlanta 2015.

My first attempt was constructed from a 32-gallon round plastic trash can. It worked great but the "boom" was lacking — it just wasn't big enough. I wanted a gigantic vortex air cannon, but one that was easy to transport (inside my car), easy to set up and break down, and inexpensive to build. To achieve this, I used PVC pipe as the modular frame, wrapped in a skin of common blue poly tarp.

The result is an intimidating, collapsible air cannon that shoots beautiful smoke rings across long distances. *Boom!*

THE STRUCTURE

The frame of the cannon is constructed from ¾" PVC pipe and consists of 2 octagons, one large and one small, connected by 8 removable lengths of pipe that act as ribs.

To create the small octagon at the front of the cannon, connect eight 15" lengths of PVC pipe with 45° elbows, and glue in place. Once dried, cut each pipe section exactly in half and add a tee fitting, facing up, to reattach each cut point — but don't glue these. Build the large octagon identically, using 2' sections of ¾" pipe.

Now join the 2 octagons together. Place the large octagon on the floor with the tee fittings facing up and insert eight 5'-long pipes (we'll call them ribs) into these fittings. The ribs should now be standing up on their own.

Position the small octagon on top of the ribs, matching up the openings of the 8 tee fittings with the 8 ribs. To do this, slightly rotate all the tee fittings on both the large and small octagons so the 8 ribs are straight and fit easily. Once the 8 ribs and fittings are lined up, glue the tee fittings in place on both octagons, being careful not to glue the tee fittings to the ribs.

To secure the ribs to the octagons, mark and drill each tee fitting where the ribs connect, allowing for a metal safety pin to pass through and hold the pieces in place. This is also an opportunity to mark numbers on the ribs and fittings (Figure Ⓐ) to easily match them up later.

To disassemble and transport the large octagon, cut it into 2 halves that can be mended back together using straight PVC coupler fittings. Note the red tape in Figure Ⓑ, indicating where the large octagon breaks apart.

To create the aperture in the small octagon for the smoke rings, outline and cut a piece of plywood and predrill (to prevent cracking)

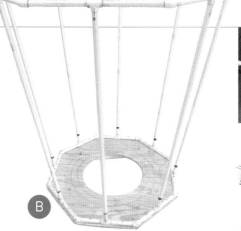
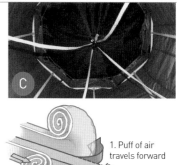

pilot holes through the PVC and plywood for drywall screws. Before attaching the plywood, draw a 20½"-diameter circle in the center of the plywood and cut it out using a jigsaw. Add 4 equally spaced eyebolts around the inside face of the plywood to act as anchors for the interior bungee cords.

THE LEGS

The front of the cannon stands on 2 thick, removable PVC pipes 1½" in diameter and 5' long. The legs fit into 2 sleeves permanently bolted to the front of the plywood with 2 bolts. The sleeves are made from 2'-long sections of 2"-diameter PVC with a straight coupler glued to the end of each.

The bolt connecting the sleeve at the top runs through both the sleeve and the plywood and acts as a hard stop for the legs so they don't slip through the top. The bolt at the bottom runs into the plywood only, through the inside of the coupler, allowing the leg to pass through. The entire leg system is non-critical, so feel free to devise your own scheme.

THE SKIN

To create the outer skin, place the cannon on the ground with the small octagon facing up. Wrap the tarp around the outside of the cannon to create a cylinder shape and pin in place using sewing pins (lots of them).

Take up the excess material by creasing and pinning pleats between each rib (8 total). There's no need to be exact about it, but the goal is to keep the skin snug around the PVC structure. It does not have to be super tight.

Once you have everything pinned up, sew the seams using heavy-duty nylon thread and heavy-duty sewing needles, then remove the pins.

At the top (smaller) opening of the tarp, fold it over and sew a seam for a nylon drawstring. This will be used to tighten the front so air won't escape around the edges.

To cinch the tarp over the larger opening, sew strips of flat nylon webbing at even intervals around the edge. Then, using a soldering iron, burn a hole through each section of webbing and thread a nylon rope though to cinch it up.

THE MEMBRANE

The membrane that covers the back of the cannon is made of waterproof coated nylon fabric. To create the membrane's giant circle, cut 2 long rectangles of fabric, reposition them into a square shape, and then join them with a strong double seam. After cutting the square down to a circle, burn the exposed edges with a lighter to prevent fraying.

To make the membrane slip over the end of the cannon like a giant shower cap, create a seam around its edge and sew in a length of nylon parachute cord. Then sew 12" of nylon webbing to the middle of the membrane to act as a pull handle. On the opposite side, sew on nylon webbing as attachment points for the interior bungee cords.

Finish up by connecting 4 bungee cords between the eye bolts on the plywood front and the attachment points on the membrane (Figure C). Place the membrane on the back, cinch everything up, and your giant vortex air cannon is ready for action!

GET VORTICAL!

To launch giant vortices of air (Figure D), pull back on the membrane handle and release it. *Boom!* It helps to have a friend stabilize the cannon as you pull. Use a Halloween fog (smoke) machine to fill the inside of the cannon with fog and you'll be able to see huge smoke rings flying through the air. ◯

See video of the giant vortex cannon and follow the complete step-by-step instructions at makezine.com/go/big-vortex.

1. Puff of air travels forward
2. Low pressure aft forms toroid vortex

Materials

- PVC pipe, ¾", 66' total length, with 45° elbows (16), tees (16), and couplers (2) for the frame
- PVC pipe, 1½", 10' total length for the legs
- PVC pipe, 2", 2' total length, with couplers (2) for sleeves
- PVC primer and cement
- Polyethylene tarp, 8'×18'
- Waterproof fabric, 5'×18'
- Nylon webbing, 1" wide
- Metal safety pins, 2½"
- Plywood sheet, ⅝"×4'×4'
- Bungee cords (4)
- Drywall screws
- Eye bolts, with washers and lock nuts (4)
- Bolts, ⅜", with washers and lock nuts (4)
- Nylon parachute cord (50')
- Fog machine

Tools

- PVC pipe cutting tool
- Cordless drill
- Sewing machine and supplies
- Measuring tape
- Jigsaw, hacksaw, and handheld circular saw
- Sander
- High-speed rotary tool
- Soldering iron
- Adjustable wrench

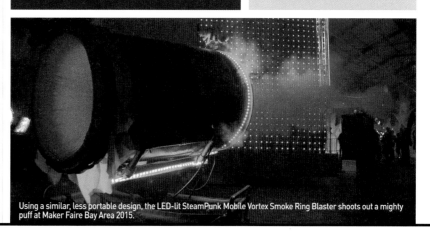

Using a similar, less portable design, the LED-lit SteamPunk Mobile Vortex Smoke Ring Blaster shoots out a mighty puff at Maker Faire Bay Area 2015.

PROJECTS | Amateur Scientist

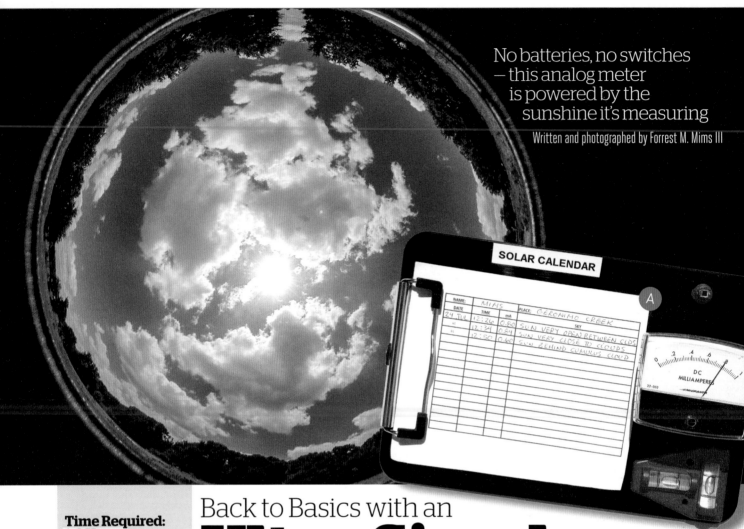

No batteries, no switches — this analog meter is powered by the sunshine it's measuring

Written and photographed by Forrest M. Mims III

Back to Basics with an
Ultra-Simple Solar Radiometer

Time Required:
1–2 Hours
Cost:
$20–$30

Materials
- **Analog panel meter, 0mA–1mA** such as All Electronics part #PMD-1MA, allelectronics.com
- **Resistor, 550Ω**
- **Photodiode, silicon, BPW34S type** available from jameco.com, digikey.com, etc.
- **Audio plug and jack, RCA or phono type** (optional)
- **Mini-clipboard or other mounting surface**
- **Bubble level**
- **Solder lugs (2)** for meter connection
- **Hookup wire**

Tools
- Drill
- Soldering iron and solder

FORREST M. MIMS III (forrestmims.org), an amateur scientist and Rolex Award winner, was named by *Discover* magazine as one of the "50 Best Brains in Science." His books have sold over 7 million copies.

SUNLIGHT PLAYS HUGE ROLES IN LIFE AS WE KNOW IT. The red and blue wavelengths of sunlight trigger the photosynthesis that is essential for the growth of plants. The tilt of Earth's axis causes the seasons: It's summer in the hemisphere tilted toward the sun and winter in the hemisphere tilted away from the sun.

I've been tracking the daily intensity of sunlight at noon since September 1989 using a variety of homemade instruments. While my instruments work well, they require batteries and power switches, both of which eventually need to be replaced. Here we will build the ultra-simple radiometer shown in Figure Ⓐ that needs no power switch or battery, for it's powered by the sunlight it measures. This radiometer will teach you much about the daily cycle of sunlight and how it's modulated by the seasons, clouds, and air pollution.

HOW IT WORKS

For many years photographers used light meters that employed a selenium photocell connected to an analog meter. The spectral response of selenium closely resembles that of the color response of the human eye. The sunlight sensor for our radiometer is a silicon photodiode, which is actually a miniature solar cell. While its spectral response peaks in the near infrared, it is far more sensitive than selenium photocells. As shown in Figure

B, the cathode pin of the photodiode is connected directly to the negative terminal of a 0–1 milliampere (mA) analog panel meter. The anode pin of the photodiode is connected to the positive terminal of the meter through a 550Ω or similar resistor that reduces the peak photodiode's current to less than 1mA in full sunlight.

While analog panel meters are considered old-fashioned, the one used here is key to the simplicity of the radiometer. Analog meters lack the resolution of digital readouts, and they're bulkier. Yet analog panel meters have a very long lifetime and other significant advantages over their digital counterparts. In this project the meter requires no battery, because it's powered by the light sensor. And the swinging needle of a panel meter shows changes and trends that are immediately recognizable, unlike the flickering numbers displayed by a digital readout.

As for reliability, decades ago I built an analog radiometer much like the one described here. Its purpose was to measure the power of the beam emitted by near-infrared LEDs. That instrument works as well today as it did when I built it in 1970, and the version described here should have an equally long life.

BUILD YOUR RADIOMETER

The radiometer shown in Figure A was assembled on a mini-clipboard to provide space for a log sheet and a bubble level. You can also use a standard-size clipboard or a piece of thin plywood or rigid plastic. To follow the design shown in Figure A, bore a ¼" hole in the board for the audio jack, above where the meter will be mounted. Place your meter over the board with its top edge centered under and at least 1" below the audio jack hole. Mark the position of the meter's 2 screw terminals. Measure the diameter of the screw terminals (the ones on mine are ⅛") and bore holes for them.

Remove the nuts from the meter's screw terminals, and insert the terminals into the holes you drilled. On the back of the board, place a solder lug over each terminal, and secure the meter and lugs in place with the nuts as shown in Figure B.

Solder a short length of connection wire between the negative meter lug and the audio jack lug. Solder a 550Ω resistor between the positive meter lug and the

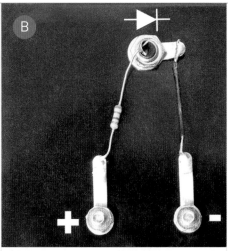

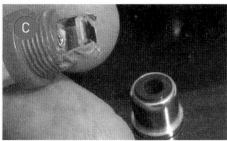

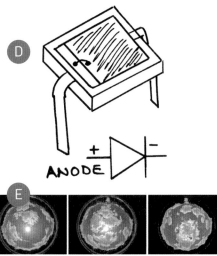

12:26 PM – 0.80mA   12:34 PM – 0.84mA   12:50 PM – 0.60mA

audio jack's center terminal. (You may need to experiment with the best resistor value if you use a photodiode other than the one specified.)

Finally, install the photodiode. You can install it directly on the clipboard, but I've installed it in an audio plug to form an interchangeable sunlight probe as shown in Figure C. For the latter method, remove the audio plug's plastic cover. Next, spread the photodiode pins outward and carefully inspect the upper side of the photodiode. A tiny wire bonded to the top surface of the silicon wafer goes to the anode (+) pin

(Figure D). Thread the photodiode pins through the connection holes in the plug's 2 terminals.

IMPORTANT:
The anode pin should be inserted through the audio plug's center terminal.

Carefully solder the photodiode pins to their respective terminals. After the connections cool, clip off the ends of the pins extending beyond the terminals.

USING THE RADIOMETER

Test the radiometer by pointing a flashlight at the sunlight probe. The meter needle should move slightly. Take the radiometer outside during daylight, and the needle will move much more. If it exceeds 1mA, you'll need to increase the resistance of the resistor. For serious studies, use a bubble level to keep the instrument level during measurements. Be sure to place the radiometer opposite the shadow formed by your head.

Figure E shows fisheye photos of the sky for 3 different summer cloud conditions. Note that the highest current occurred when the sun was closely surrounded by clouds. Also note the surprisingly high current when the sun was blocked by a large cloud.

You can produce serious data by measuring sunlight at solar noon for a full year. I've done this at my Texas site for more than 26 years using a variety of DIY radiometers, and the graphed data shows an obvious seasonal cycle. Clouds, dust, and smoke can greatly reduce sunlight. The thickest haze in Texas is caused by Saharan dust from Africa and power plant haze from the Ohio Valley during summer and fall. Smoke from distant forest fires and agricultural burning in Mexico also causes thick haze.

GOING FURTHER

» To study daily trends in the data and the effects of clouds, you can make a video or elapsed-time photos of the meter.
» Use the radiometer to monitor the red wavelengths of sunlight that stimulate photosynthesis by placing a red filter over the photodiode.

See the graphed data from Texas at makezine.com/go/solar-radiometer.

PROJECTS

1+2+3 Wearable EL Flame

Written and photographed by Helga Hansen

ELECTROLUMINESCENT LIGHTING (EL) CREATES BEAUTIFUL EFFECTS WITH VERY LITTLE EFFORT. You can spice up your apparel with glowing EL wire accents, or use flat, flexible EL panels to cut larger custom shapes. Follow along to shape an EL panel into an eye-catching, wearable flame.

1. APPLY DESIGN

Draw directly onto the back of the panel. We recommend practicing on scrap paper first, cutting out the best design, then tracing it directly onto the panel. Your design must be a continuous piece and incorporate the power connector.

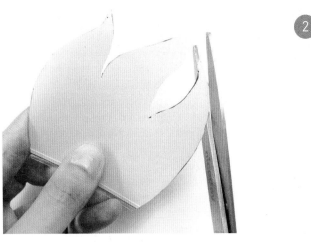

TIP: Before starting, test to make sure that the EL panel lights up. If it doesn't, try cutting away the heat-shrink and reflowing the solder connections between the panel and the wires.

2. CUT IT OUT

Trim the design out of your EL panel using scissors. While cutting, the lamination may rise around the edges a bit. This shouldn't affect the panel's performance and you can use a razor or hobby knife to trim the peeling lamination. Be careful not to damage or cut through the power connection.

3. SEAL EDGES AND ATTACH PIN

To seal the EL panel's lamination and protect against shorting out, apply clear nail polish to the edges of the design. Wipe away any marker from the back before applying or the polish may dissolve and smear the ink. Alternatively, you could wrap the design in packing tape or laminate.

Place batteries into the inverter pack and connect it to the panel cable. Attach a pin backing with hot glue and let it cool. Now rock your hot, new accessory! ◉

HELGA HANSEN is an editor at the German edition of *Make:*. She loves sci-fi and bow ties, and fixes perler beads with a soldering iron.

Time Required: 30 Minutes
Cost: $20–$30

You will need:
» EL panel, 10cm×10cm SparkFun.com
» EL inverter battery pack
» Batteries
» Pin backing
» Scissors
» Hobby knife
» Marker
» Clear nail polish or tape
» Hot glue gun and glue

SCIENTIFICS® DIRECT

Trusted, Inspired, Educational & Fun
1-800-818-4955 • ScientificsOnline.com

FREE GROUND SHIPPING on orders $75+
USE **MAKEMAG**

Your Trusted S.T.E.M. Resource Center for Innovation & Quality

At **Scientifics Direct**, we have hundreds of science, technology, engineering products that ignite the imagination and help make learning fun. We specialize in hands-on, applied science kits designed specifically for learners age 8 to adult.

From our new **Hydraulic Arm Robot Kit** with six axes of detailed movements to our **Duino Programming Kits** with sturdy metal carry cases, we have S.T.E.M. related product solutions for every need and budget.

Visit us today at
ScientificsOnline.com

Duino Essentials Learning Kit......#3155100
DuinoKit Jr......#3155318

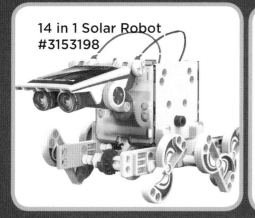

14 in 1 Solar Robot
#3153198

IQ Key Perfect 600
#3155217

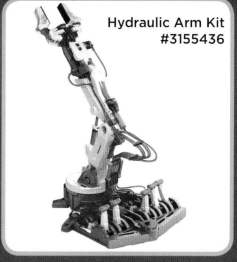

Hydraulic Arm Kit
#3155436

FIND HUNDREDS OF OTHER UNIQUE SCIENTIFIC ACTIVITIES, GADGETS, COLLECTIBLES, AND MORE.
SCIENTIFICSONLINE.COM

*Free ground shipping valid through 07/31/16

TOOLBOX
GADGETS AND GEAR FOR MAKERS
Tell us about your faves: *editor@makezine.com*

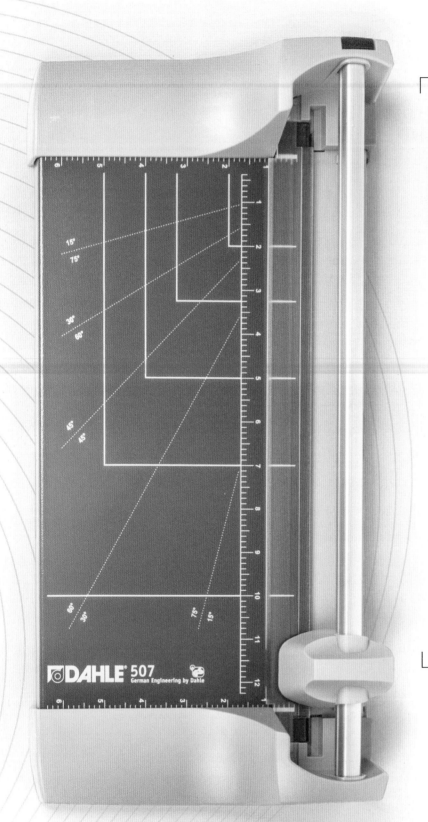

Dahle 507 Personal Rolling Trimmer
$79 dahle.com

When looking for a personal, high-volume trimmer, the Dahle 507 is a great choice. The aluminum base is far sturdier than the usual plastic type, and the greater weight helps ensure stability for clean cuts. Over time, a self-sharpening blade saves you money on replacements, and the blade mechanism is fully enclosed for a safer system.

The Dahle trimmer easily handles several kinds of paper stock, from normal printer weight up to cardstock and photo paper. It's incredibly easy to do both short and long cuts with the gliding blade action. This trimmer doesn't sacrifice cut quality either — edges are clean and crisp, even when slicing through multiple sheets of cardstock. The easy-to-use guides mean your lines are straight and your cuts are exactly the right size. A clear, automatic clamp keeps your papers right where you want them while making it easy to see where the cut line is. Well-constructed, this rolling trimmer will last you for years to come.

— *Hep Svadja*

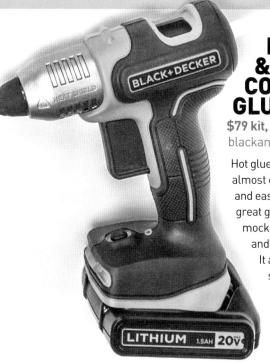

BLACK & DECKER CORDLESS GLUE GUN

$79 kit, $39 bare tool
blackanddecker.com

Hot glue is a necessity for almost every Maker. It is fast and easy to use, making it a great go-to tool for cardboard mock-ups, temporary fixtures, and permanent joints alike. It allows you to create a strong bond on almost any material, and the bond doesn't have to be permanent unless you want it to be.

Glue gun technology has basically been the same since its conception: a simple plastic gun with a long cord that always seems to get in the way. The Black and Decker BDCGG20 is a 20V lithium ion battery-powered cordless glue gun. Not having the cord allows you to have greater control and increases its portability. It feels more like a quality power drill than a glue gun.

It takes the battery about 90 seconds to heat up the glue and will give you around 3 hours of run time. The glue gun will set you back about $79 with a battery and charger, or $39 without.

— Dan Maxey

DEWALT DWE6423K 5" RANDOM ORBIT SANDER

$80 dewalt.com

The DeWalt DWE6423K orbital sander focuses on reducing dust while also utilizing a counterweight to lower the overall vibration of the tool, for a more comfortable and less fatiguing user experience. Out of the box it looks a bit different from its predecessors: The overmold handle is smaller for better hand control, and they've also included a rubber skirt around the sanding surface.

The tool is fairly lightweight at just 2.9lbs, which is pretty great considering it has a 3-amp motor with variable speed. I tested it side-by-side against my older DeWalt D26451 orbital sander, and could immediately feel it was much smoother and not jarring.

It's obvious they took extra care in dust collection with the new model. Not only does it come with a bag, I was able to connect it to my shop vacuum with minimal dust escaping. Overall, I'm impressed that they delivered on their promises. This is a mighty sander in a small, thoughtful package.

— Emily Coker

TEKTON 26759 SLOTTED AND PHILLIPS SCREWDRIVER SET, 16-PIECE

$50 tektontools.com

These aren't your typical screwdrivers — they're better. Tekton's USA-made screwdrivers are as durable as the old classics, but offer high torque action with a new level of comfort.

At first sight, I was concerned by the unusual 3-sided handle design, thinking it might be uncomfortable, but this feeling slowly faded into intrigue. After repeated use, I had no hand or wrist fatigue. In fact, the clever design allowed for better control and easy finger spinning.

In addition to the improved handle, the screwdrivers are made of chrome molybdenum steel for high strength, and finished in a hard black oxide for corrosion resistance.

Overall, this set is affordable, ergonomic, and built to last.

— EC

TOOLBOX

HOBBY CREEK THIRD HAND
$45 hobbycreek.com

This helping hand unit is a pretty fancy little setup with four work-holding arms to help get the job done. The heavy aluminum milled base offers stability, and provides storage for small parts while working — a nice touch! The well-built arms are stiff yet flexible, so you can move them to perfect placement.

One major concern is the heat-shrink material on the alligator clips — when you're applying solder directly to non-insulated wires, the plastic just melts away. I would suggest removing the casing while doing direct hot work.

We spoke to Hobby Creek about this, and they said that they are switching from heat-shrink to medical-grade silicone tubing covers as standard on all their products. The silicone tubing covers should be more robust and resilient to heat transfer.

This tool is great for most applications, especially steadying intricate PCBs for soldering.

— EC

XYZROBOT BOLIDE Y-01 ADVANCED HUMANOID ROBOT
$750 (fully assembled) xyzrobot.com

The latest to hit the market in STEM robotics is the Bolide Y-01 Advanced Humanoid Robot by XYZrobot. The sturdy bot stands 16" tall and is ready to go right out of the box. I chose the fully assembled version (which comes with a higher price tag), but it's also sold in both semi-assembled and DIY packages.

The Bolide comes with preset commands like doing push ups, dancing, and waving. It can be controlled several ways: with the provided remote control, on your phone via BLE 4.0 with the companion app, or via an SD Card with four programmable buttons on the back of the bot. The Bolide is built with 18 smart servos that can detect temperature, speed, and position, as well as an ATmega 1208 processor, internal IR sensor for distance, controllable RGB LEDs, speakers, and an accelerometer to pick itself up.

XYZrobot has downloads available on their site for 3D printing your own replacements or unique parts. You can further customize your robot with the XYZrobot editor that runs off the Arduino 1.0.6 IDE. Overall, the Bolide is well-built and covers several areas of hands-on learning. This would be a perfect addition to a classroom or library where it can engage a larger audience and offset the cost.

— EC

SPEEDY STITCHER SEWING AWL

$12–$30 speedystitcher.com

This is the perfect tool for those heavy-duty sewing jobs, especially if you don't own an industrial sewing machine. People have been using this handy awl since 1909, which speaks to its reliability. The stitcher comes with one straight needle, one curved needle, and a spool of waxed polyester thread with 52-pound tensile strength. The instructions are easy to understand and the awl is ready to use out of the box.

The stitcher creates strong lock stitches that can punch through almost any material. You can also purchase additional needle sizes, thread, and other useful goodies. This is a welcome addition to my toolbox — I'm just sad I didn't know about it sooner.

— EC

ADAFRUIT NEOPIXEL JEWEL RGBW LED

$7 for Natural White Jewel
adafruit.com

Adafruit's new Neopixel RGBW LED products offer a number of advantages over their standard RGB Neopixels, and not just for those times when you want a purer white light.

To start, the white LED is separate and independently controllable. This means that you can achieve white light without having to turn all three red, green, and blue channels up, which is simpler code-wise and should result in lower power draw. Plus, with a separate white LED element, you have your choice of cool, warm, or natural white color temperatures. Cooler light is whiter (sometimes bluer), warmer will be yellower, and natural will be in between with slight neutral yellow tones. You have to make your choice at the time of purchase; I prefer neutral white.

The new RGBW Neopixels have another, less obvious benefit: built-in diffusion. Each 5050-sized LED element has a translucent lens, which results in more uniform light spread and better color mixing. The difference is slight, but noticeable — although you'll still want your own board-wide diffusion to achieve certain lighting looks.

The RGBW Neopixel products cost a little more: each of the 7-LED RGBW Jewels is priced at $6.95, and the RGB version is $5.95. But even if you don't plan on using the white LED channel, the built-in diffusion lens is reason enough to opt for RGBW Neopixel over RGB.

— Stuart Deutsch

STAEDTLER SILVER SERIES MECHANICAL PENCILS

$11–$13 staedtler.us

In a pencil, function trumps form, but the Staedtler line doesn't sacrifice either. Its precision-engraved grip and smooth barrel look sleek enough for the office while still being robust enough for the workshop. The feeling in your hand is solid, while the no-slip grip ensures that it will stay there, and the sturdy but removable clip keeps it hooked in place.

Available in plenty of point sizes from 0.3mm extra fine up to 2.0mm bold, there is a lead option to fit every drawing or drafting style. I like to use a 2.0 for rough concepts, and I have a 0.5 for detail work. All models have a tiny eraser under the cap except the 2.0mm pencil, which can fit a normal pencil eraser head over the end cap instead. Lead stick refills are available at most office supply stores and online, and in plenty of different graphite grades, making this already great pencil even more versatile.

— HS

BOOKS

MADE BY RAFFI

by Craig Pomranz
$17 craigpomranz.com

As adults, it can be much easier for us to celebrate what makes us unique than it is for kids to accept what makes them different from their peers. In his new children's book *Made by Raffi*, Craig Pomranz tells the story of a young boy who, after worrying that he's not like "normal" boys, learns to be comfortable with himself through his newfound love of knitting. If you know a young Maker who is struggling with the awesome burden of being creative, they might find solace in Raffi's story. In addition to celebrating your unique hobbies and identity, the story teaches a great lesson about gender stereotypes, and has instructions for making a pretty cool cape, too.

— Sophia Smith

TOOLBOX
3D Printer Review

ULTIMAKER 2+

A NEW AND IMPROVED EXTRUDER PRODUCES CRISP, CLEAN, AND IMPRESSIVE PRINTS
WRITTEN BY MATT STULTZ

PRINT SCORES 0 1 2 3 4 5

- Vertical Surface Finish
- Horizontal Finish
- Dimensional Accuracy
- Overhangs
- Bridging
- Negative Space
- Retraction
- Support Material (pass)
- Z Wobble (pass)

TOTAL 34

MANUFACTURER Ultimaker
PRICE AS TESTED $2,500
BUILD VOLUME 223mm×223mm×205mm
BED STYLE Heated glass
FILAMENT SIZE 3mm
OPEN FILAMENT? Yes
TEMPERATURE CONTROL? Yes, tool head (180–260°C), bed (50–100°C)
PRINT UNTETHERED? Yes (SD card)
ONBOARD CONTROLS? Yes (LCD with control wheel)
HOST/SLICER SOFTWARE Cura
OS Linux, Mac, Windows
FIRMWARE Open Marlin
OPEN SOFTWARE? Yes, both software and firmware
OPEN HARDWARE? Yes, Creative Commons Attribution-NonCommercial 3.0
MAXIMUM DECIBELS 76.8

OVER THE PAST 5 YEARS ULTIMAKER HAS BEEN BUILDING MACHINES THAT PUSH THE ENVELOPE OF PRINT QUALITY FOR DESKTOP 3D PRINTERS, garnering them a community ready to sing their praises. The Ultimaker 2 brought a new look to the series that helped attract a wider audience, but it also brought a problem in its finicky extrusion system. The + in the Ultimaker 2+ is all about one thing: a brand new extrusion system, and it makes all the difference!

IF IT AIN'T BROKE DON'T FIX IT
At first glance, the only distinguishable difference between the 2+ and its predecessor is the + sign on the front of the machine. The machines are the same size, with the same build area and control panels. Look at the extruder, however, and you will notice the upgrades. The popular Olsson block kit (a third-party upgrade to Ultimaker 2 machines that Ultimaker has begun reselling) comes installed, allowing quick nozzle changes between the included .25, .4, .6, and .8mm sizes. The .4mm is installed by default. New fan shrouds create more even cooling coverage to help lock in the fine details of your print.

The really exciting changes are found on the back of the machine. The new feeder improves grip and provides more torque to push the filament through the Bowden tube. Removing or manually feeding filament is no longer an issue either — a grip-release button allows the user to hand feed or pull out the filament without the need for the motor. The new feeder also eliminates filament skipping in the extruder motor, which caused a tell-tale clicking sound associated with the Ultimaker 2.

A RESOLUTION REVOLUTION
The prints are what you expect from an Ultimaker: crisp, clean, and incredibly impressive. One of the core principles of *Make:*'s 3D printer testing plan is that we use the medium default settings when reviewing the machines. Most printers have that default layer height at about .2mm, but Ultimaker printers halve that at .1mm, resulting in prints that automatically look better than their competitors'. Ultimakers can go even further with their fine settings, down to a resolution of .06mm. Of course, everything comes with a trade off, and fine resolution means longer print times. Our astronaut print (which we try to scale to an 8-hour print time) is around half the size of those printed on many other machines.

While in our tests the Overhang and Bridging scores are good, these are the two areas that the Ultimaker 2+ could still use some improvement. This shows that while the new fan ducts did help with more even airflow, more cooling could still help the 2+.

CONCLUSION
If you are a user who already has an Ultimaker 2 or 2 Extended and are having buyer's remorse while reading this, don't worry — Ultimaker has you covered. They have recently launched a drop-in replacement kit, allowing for a quick upgrade.

Ultimaker's reputation for superior quality 3D prints is well deserved. The 2+ shows that the company is willing to look at their problem areas and listen to customer feedback to make their machines even better.

PRO TIPS
While not a perfect solution, you can change between 1.75mm and 3mm filament by setting the diameter in the filament profile on the machine. I copied the 3mm profile and created a new one called PLA175 to make switching back and forth easy.

WHY TO BUY
The Ultimaker 2+ gives its user superior print quality while still maintaining an open environment. And you'll be the envy of all of your 3D printing friends!

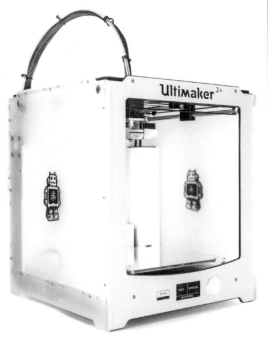

RESULTS

Matt Stultz is the 3D Printing and Digital Fabrication lead for *Make:*. He is also the founder and organizer of 3DPPVD and Ocean State Maker Mill, where he spends his time tinkering in Rhode Island.

For more reviews and testing procedures, go to makezine.com/go/3dp-comparison.

Make: Marketplace

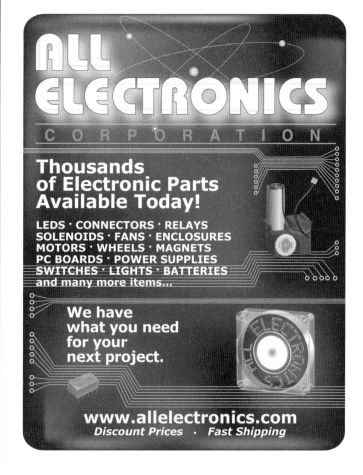

Make: Marketplace

NEW BOOKS — Maker Shed

makershed.com

MAKE: FIRE From portable fire pit to workshop foundry to "booshing" flame canon, this is your complete, hands-on guide to working with propane. With six fully illustrated projects, you'll learn the science of propane and discover the equipment and parts that allow you to work with it — including Arduino.

DIY DRONE AND QUADCOPTER PROJECTS FROM THE PAGES OF MAKE: The best tutorials, projects, and commentary on drones from the pages of Make: magazine are collected here to help you learn, build, and prepare for lift-off. Tackle state-of-the-art projects using fully illustrated, step-by-step instructions that show you how to build different drones from scratch.

MAKE: FUN! Professional toy inventor Bob Knetzger shares his secrets for turning everyday materials into toys, games, and clever amusements. This lavishly illustrated book is packed with more than 40 projects that show you how to work with simple electronics, mold and sculpt plastics, create your own metal forge, build toys that demonstrate scientific principles, and more.

BIOSENSING FOR EVERYBODY
shop.openbci.com

10% Off
All OpenBCI Gear
Discount Code:
MakeOpenBCI

OpenBCI

PLASTIC INJECTION MOLDING MACHINES - STARTING AT $595!

Mold your own plastic parts. Perfect for inventors, schools or companies. For details & videos visit our webpage.

The affordable Model 20A turns your workshop drill press into an efficient plastic injection molding machine. Simple to operate and it includes a digital temperature controller.

No expensive tooling is required - use aluminum or epoxy molds.

The bench Model 150A features a larger shot capacity and is perfect for protoyping or short production runs. Capable of producing up to 180 parts per hour!

We also carry:
* MOLDS
* CLAMPS
* ACCESSORIES
* PLASTIC PELLETS

MADE IN THE USA

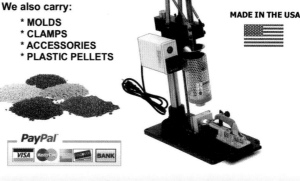

PayPal — VISA MasterCard Discover BANK

www.techkits.com   707-328-6244

Introducing Make: Money

Sell MAKE subscriptions and make money for your organization. Your group keeps 50% of every subscription you sell!

makezine.com/makemoneyprogram

CadSoft EAGLE

Layout Editor. Schematic Editor. Autorouter.

Full EAGLE functionality at a fraction of the cost.

Our EAGLE Make Portfolio for you:

Make Personal

Make Pro

VISIT
www.makeeagle.com
to learn more.

EAGLE

MAKE. Create. Innovate.

Make: OVER THE TOP

Crazy Train:

Written by James Burke

Ozzy Osbourne joins the Gremlins Carnival Club

[Verse]
Gremlins, that's part of the show
Four generations building in droves
Maybe. Lights will be great
To burn a megawatt yet need higher rates

[Chorus]
LED lamps gleaming
The generator's to blame
I'm leadin' a parade on a crazy train
I'm makin' sure our lorry stays in the lane

[Verse]
We've partied as winners
We've partied like fools
We've worked with other float clubs
Who hacked their own tools
Once were RAF blokes who founded the role
Now it's all families that are in control

[Chorus]
Every year we're scheming
With themes that are insane
I'm leadin' a parade on a crazy train
I'm makin' sure our lorry stays in the lane

[Bridge]
I know that we've won over two hundred times
From Bridgwater to Midsomer, yeah yeah

[Verse]
Heirs of some bad scores
Complacency's not one
Overcoming troubles
Until the job gets done
Crazy, the costumes we wear
We're now planning the new float for the next year

[Chorus]
Townsfolk out there cheering
Not hard to explain
I'm leadin' a parade on a crazy train
I'm makin' sure our lorry stays in the lane ◐

Founded in 1948, the Gremlins Carnival Club is an award-winning float building group that has enchanted carnival-goers with insanely bright light displays all over the U.K.